DeWALT®

CONTRACTOR'S FORMS & LETTERS

WITH INTERACTIVE CD

By Paul Rosenberg

Published by:

 DELMAR
CENGAGE Learning™

www.DeWALT.com/guides

DeWALT Contractor's Forms & Letters
Paul Rosenberg

Vice President, Technology and Trades Professional Business Unit: . . . . Gregory L. Clayton
Product Development Manager: . . . . Robert Person
Executive Marketing Manager: . . . . Taryn Zlatin
Marketing Manager: . . . . Marissa Maiella

For product information and technology assistance, contact us at **Professional Group Cengage Learning Customer & Sales Support, 1-800-354-9706.**
For permission to use material from this text or product, submit all requests online at **cengage.com/permissions**.
Further permissions questions can be e-mailed to **permissionrequest@cengage.com**.

ISBN-13: 978-0-9777183-2-0
ISBN-10: 0-9777183-2-8

Delmar
5 Maxwell Drive
Clifton Park, NY 12065-2919
USA

Cengage Learning is a leading provider of customized learning solutions with office locations around the globe, including Singapore, the United Kingdom, Australia, Mexico, Brazil and Japan. Locate your local office at: **international.cengage.com/region**.
Cengage Learning products are represented in Canada by Nelson Education, Ltd.

For your lifelong learning solutions, visit **delmar.cengage.com**.
Visit our corporate website at **cengage.com**.

Notice to the Reader
Information contained in this work has been obtained from sources believed to be reliable. However, neither DeWALT, Cengage Learning nor its authors guarantee the accuracy or completeness of any information published herein, and neither DeWALT, Cengage Learning nor its authors shall be responsible for any errors, omissions, damages, liabilities, or personal injuries arising out of use of this information. This work is published with the understanding that DeWALT, Cengage Learning and its authors are supplying information but are not attempting to render engineering or other professional services. If such services are required, the assistance of an appropriate professional should be sought. The reader is expressly warned to consider and adopt all safety precautions and to avoid all potential hazards. The publisher and DeWALT make no representation or warranties of any kind, nor are any such representations implied with respect to the material set forth here. Neither the publisher nor DeWALT shall be liable for any special, consequential, or exemplary damages resulting, in whole or part, from the readers' use of, or reliance upon, this material.

Printed in Canada

2 3 4 5 XX 10 09 08

NOTE TO THE READER

Although great care and effort has gone into the development of the forms, letters and contracts contained within this book/CD-ROM, we cannot guarantee that all of the information is always correct, complete, up-to-date, or appropriate for your specific needs. These forms, letters and contracts may have to be modified to suit your needs and state law requirements. An attorney should be consulted for all legal matters.

Furthermore, we cannot guarantee that the pre-programmed forms do not contain any miscalculations or mathematical errors. Please double-check all calculations before submitting any proposals.

This information is given with the understanding that neither DeWALT, the publisher, author, or seller is engaged in rendering any legal, business or financial advice to the purchaser or to the general public.

In no event will we (Delmar Cengage Learning), DeWALT, our agents, partners, or affiliates be responsible for any direct, indirect, incidental, or consequential damages (including, but not limited to, procurement of substitute goods or services; loss of use, data, or profits; or business interruption) however caused and based on any theory of liability, whether in contract, strict liability, or tort (including negligence or otherwise) arising in any way out of the use of the content provided within this book/CD-ROM.

ABOUT THE AUTHOR

Paul Rosenberg has an extensive background in the construction, data, electrical, HVAC, and plumbing trades. He is a leading voice in the electrical industry with years of experience from apprentice to project manager. Paul has written for all of the leading electrical industry and low-voltage industry magazines and has authored more than 50 books.

In addition, he wrote the first standard for the installation of optical cables (ANSI-NEIS-301) was awarded a patent for a power transmission module, and both developed and taught nineteen courses for Iowa State University (College of Engineering).

Paul is a regular contributor to Power Outlet Magazine, and works as an expert witness in legal cases. He has performed consulting work for NASA, the National Electrical Contractors Association, the National Association of Electrical Distributors, the Independent Electrical Contractors Association, the Mechanical Contractors Association of America, Westinghouse Electric, Wal-Mart Stores, and many others.

Paul speaks occasionally at industry events.

CONTENTS

DeWALT CONTRACTOR'S FORMS & LETTERS

by
Paul Rosenberg

CORRESPONDENCE AND BUSINESS IN GENERAL

This book contains pertinent information on business correspondence for construction businesses.

Method of Sending

It used to be that all correspondence was done on paper, and even though e-mail is rapidly taking over that role, you should still use paper for important documents. (Although, for extremely important documents, encrypted and digitally-signed e-mails are best.) If you use e-mail for any serious correspondence, make sure that you save a copy to a separate folder, and be sure that you will be able to find it later.

Contracts

To their own disadvantage, most construction companies (especially small construction companies) do not spend enough time and effort reviewing contracts.

Contracts can be dull and difficult. Nonetheless, you are likely to get yourself in trouble if you do not pay attention to the details. (Regarding contracts, the old phrase is true—"The devil is in the details.") Another big difficulty is that companies are afraid to demand changes in the contracts. They seem to have a fear of offending the owner by asking for fair contractual terms. Do not let this happen to you. If you are risking your money on a construction project, you should demand fair terms. Contracts almost always benefit the writer; if you are afraid to cross out bad terms, you will be walking into the situation at a disadvantage. Consider the following true statement from Robert Ringer—"The results you produce in life will be inversely proportional to the degree to which you are intimidated."

There are several different types of contract documents is this book. Some are design/build agreements, some are basic contracts and proposals, and others deal with your bids.

Customer Relations

It has been said that it is ten times harder to get a new customer than it is to keep an existing customer. Whether this ratio is correct or not, the fact is that it's far easier to keep a customer than to get a new one. Because of this, it is very important for you to do everything possible to keep your customers happy. Unfortunately, most construction companies do not do this very well.

Construction firms typically get so busy with day-to-day problems that they have no time to do any marketing work. Most of the required effort consists of occasional letters or phone calls and using well written communications as a matter of course.

The letters that you find described should be a big help along these lines. They are written to keep your customers happy and improve your image in their minds.

In all of your communications, remember that everything is marketing. Every type of communication that leaves your office (letter, phone call, e-mail) says something about your company. Every communication leaves an image in the recipient's mind. If it is a good image, your company benefits; if it is a bad impression, your company will eventually suffer. The images projected by your communications will accumulate in your customers' minds. Over the course of time, you will build an image that attracts them or one that repels them.

Employees

Since employees are the people who actually do the things that your company needs to accomplish, communicating with them is of extreme importance. Employee correspondence should be used more by contractors. Among them are employer/employee agreements, lists of responsibilities, and schedules. These forms clearly define the responsibilities of the involved parties and will limit your exposure to lawsuits by disgruntled employees. If you do not use such forms already, you should take a good look at these and consider using them.

Legal Letters and Forms

Legal letters and legal forms are important to almost all construction companies, but especially to general contractors and developers who are frequently involved with complex real estate transactions. Real estate transactions are a traditional investment business for contractors.

Projects

Project managers are notoriously pressed for time and rarely take the time to write many letters. It is not because they are being negligent in their duties; it is simply that they have no available time.

Not only is writing letters critical from a marketing standpoint, but it is critical for legal protection as well. Running a construction project is a lot like warfare; you have to protect your company's interests at all times. Most projects run fairly well, but when a construction project fails, it falls apart very quickly, and usually completely. Once it falls apart, everyone involved will be trying desperately to prove their company's innocence by laying the blame on other people. When this happens, the company with the best paper trail usually comes out as the winner. Remember that a written record of an event holds up very well in court. Remembrances of spoken statements do not.

Vendors

Most of the letters you will write to your vendors will be written by your bookkeeper or purchasing manager. These letters are not especially difficult, although there are often so many such letters to write that the bookkeeper simply cannot keep up, so matters go unresolved until there are problems.

As I have said in other places, perhaps the most important thing about business communications is that they actually get out. For construction companies, the most difficult part of getting these communications out is the shortage of time. This is why loading these letters into your bookkeeper's computer makes so much sense. With a few keystrokes, the letter will be ready to send.

EDUCATING YOURSELF, ENRICHING YOURSELF

The construction business has long been considered a great area of opportunity for a person to make money and become rich. The difficulty is that success requires the contractor to make transitions from tradesman to entrepreneur to investor. This requires continuing education.

Making money in contracting requires a good deal of specialized knowledge. You must understand supply and demand, economic forecasting, marketing, and even the psychology of your workers. The following are books and other resources that will inspire you and help you. Note that these are not treatises on accounting, law, etc. (You should understand the basics of those things, then hire professionals to do them for you.)

Here are several good books that will help you to understand what business (not just your trade) is really all about:

Whatever Happened to Justice? and *Whatever Happened to Penny Candy?*, both by Richard Maybury. These were actually written for high school students, but they are worth while reading for everyone. "*Justice*" explains law and "*Penny Candy*" explains economics. These are the best and easiest primers you will ever read.

Economics in One Lesson, by Henry Hazlitt. If you want to understand, with a minimum of pain, what economists are talking about, read this book after you finish "*Penny Candy.*"

The Richest Man in Babylon, by George Clason. This 1920s collection of simple stories is still a classic. It contains all of the real basics regarding money. Just do what this guy says, and you're on your way.

Think and Grow Rich, by Napoleon Hill. A bit quirky, but a very important work. If you get this book, you can avoid spending a lot of money on "success" seminars—they all get their material from this little book.

Million Dollar Ideas, by Robert Ringer. Ringer has produced many good entrepreneurial books, and this is one of them. This guy has been in the game, won, lost, and remembered the lessons. Don't miss him.

The Third Wave, by Alvin Toffler. Read the first ten chapters. They will help you understand the basic directions in which the world is moving. (The rest of the book is unnecessary.) For more on this subject, read *The Sovereign Individual*, by James Davidson and William Rees-Mogg. This is an exceptionally interesting and well-reasoned look at the near future. You may also want to read *The Singularity is Coming*, by Ray Kurzweil.

Atlas Shrugged, by Ayn Rand. Rand was hated in her time, and is still hated in some circles. Nonetheless, her work has inspired many people to great success. This book was rated in a Library of Congress survey, second only to the Bible, as the most inspirational book ever. Read it and find out why. Rand is definitely strident and opinionated; you don't have to agree with everything she says, but you must understand it. You will also find that this is one of the few books that will actually inspire you to go out and attack the world of commerce.

CHAPTER ONE

Forms and Letters for Your Company

This chapter contains useful forms, contracts, and letters based on your business correspondence for the construction trade.

FORMS

Company Planning Meeting Topics

This is a list of topics to be covered at an annual company meeting. Such meetings are very important and should not be neglected. This list of topics should be applicable to all construction companies.

Overhead Calculations

This is a form to assist you in calculating your annual overhead. Overhead is usually expressed as a percentage of sales. The general method of calculating overhead is to total all of the company's expenses that are not related to any project, then divide them by total sales. Overhead figures vary widely for different types of construction operations, but most companies have overhead figures in the 5–20% range.

Project Cost Data Form

This is, essentially, an estimate sheet for a construction project. It has provisions for most of the essential cost figures. You may modify this form for your specific needs.

Conditions of Employment

With our current legal situation treating a job as an entitlement, hiring people can be a hazardous situation. To alleviate some of this problem, a number of companies have instituted policies mandating employment agreements for every employee. This is a sample of such an agreement. You should send a copy of any employee agreement to your attorney before using it to make sure that he or she approves of it.

Hourly Labor Rates Form

This form is for developing your final labor prices for regular, overtime, and double-time costs. Use it to determine the actual and final cost of your employee to your company.

Project Survey Form

This is a basic form to aid in planning out the project.

Employee Evaluation Form

These are forms to evaluate your company's employees. Be sure to use the space provided for notes. Humans are very complex, and short descriptions are misleading.

Daily Time and Labor Distribution Report

This is more of an in-house form than the ones you will find in the Jobs section. Use it to see how many workers are working on any given day. It is useful for checking work-flows and cash-flows.

Employee Data Form

This is a sheet for keeping data on your employees. It should be filled out for each employee and kept in a safe place.

Employee Time Sheet

This is a weekly time sheet to track employees hours.

Emergency Contact Form

This form is to be filled out for each employee, posted in the field office, and distributed to all personnel.

Emergency Response Form

This form should be filled out and signed by each employee so they are aware of the required emergency response to hazardous substances.

Accident Report

This is a form to be filled out whenever one of your employees has any sort of accident. These forms can be very important for insurance and legal purposes. Make sure that you fill them out every time, fill them out completely, and save them in a safe place.

Safety Plan

This outlines all the safety procedures and points of contact in case of any emergency.

Employee Emergency Action Plan

This outlines the action plan in case of an emergency for all employees on construction sites.

Truck Inspection Sheet

This is a maintenance checklist and record for company vehicles. Use these regularly, not only to keep records of repairs, but to make sure that all vehicles are properly maintained.

Travel Expense Report

This is a form to track the mileage and expenses for any travel affiliated with a job.

Equipment Ledger

This is a basic record of operation for construction equipment. It is good for costing, for maintenance, and for accounting cross-verifications.

Employee Driving Record

In the current legal environment, you need to protect yourself against intrusive lawsuits. Use this form to prove that you checked on your employees and their suitability to drive company vehicles.

Cash Paid Out Receipt

Essentially, this is a receipt to be used when anyone takes cash out of the company, especially for petty cash boxes and the like.

Safety Agreement Form

This is a form to be signed by each employee, verifying that he/she has received the requisite safety information and training. With injury lawsuits being as common as they are now, keeping these documents is necessary.

LETTERS

Employee Warning Letter #1

This is a final warning letter to an employee. This letter is written regarding the poor conduct of an employee. In the letter, the employee is warned that this is the last chance he or she has; either there is immediate improvement, or termination of employment will result. The wording of this letter mentions attendance. Obviously this can be changed for differing circumstances; absenteeism was mentioned since it is the most common of severe problems. Health problems are also mentioned, with personal problems (divorces, deaths, etc.) mentioned in the parenthesis. These also can be changed for specific circumstances. The tone of the letter is (and should remain after revision) friendly, giving the feeling that you are sorry to be forced into such an extreme decision.

Employee Warning Letter #2

This is a general warning to employees regarding drugs. This is a very serious warning to all employees regarding substance abuse (liquor, drugs, etc.). It can be used, as is, in response to rumors of drinking or drug use. It could also be modified slightly and used as a general reminder of the company's policy to immediately terminate anyone found using such substances. It should be distributed in pay envelopes or by some similar method.

Termination Letter #1

Probably the most unpleasant letter to write is the termination notice. Assuming that you are not firing the person for reasons of conduct, you should make the termination notice as friendly and sympathetic as possible. Try to assure the employee that you think well of them and that you will be thinking of them should an appropriate position open up.

Termination Letter #2

This is a termination notice to a problem employee. Since the legal situation in the United States is not good for employers, this notice restates problems that the employee has caused and the reasons for termination. This will be helpful to your company if the former employee files a wrongful termination suit against you. If you have reason to expect a real problem from the employee who is being terminated, e-mail a copy of this letter to your attorney prior to giving the notice to the employee and note it on the letter.

Reporting Accidents Letter

This is a general letter to all employees regarding the reporting of on-the-job accidents. It is assumed that copies of this letter also will be stuffed in each employee's paycheck; or that some similar form of distribution will be used. Note that this letter specifies that the foremen have copies of the appropriate forms. Make sure that they do before you distribute such a letter. The letter is specially written to give the employees an incentive to fill out the forms promptly, since it tells them that failure to do so could jeopardize their claims.

Injury Letter

This is a condolence letter to an injured employee. This simple letter expresses your sorrow at the employee's injury and expresses your hope of a speedy recovery. Note that it also mentions a gift that is sent along with the letter. A gift given in such circumstances will generally secure a bit of good will toward you and your company.

Insurance Change Notice

This is a letter regarding changes necessitated by insurance. This letter informs all employees of changes in conduct that will be necessary because of insurance regulations. It leaves you a blank place for filling in the details of the changes, and also clearly informs the employee that he or she must comply with the changes. It is assumed that copies of this letter will be stuffed in each employee's paycheck, or that some similar form of distribution will be used.

Letter of Appreciation

A big part of a workers' motivation has to do with what goes on in his/her family life. And even though you have no say in your employees' personal lives, you can certainly do something about how their families feel about your company. This is a personal letter from the owner or manager of the company to the worker's spouse. The purpose of the letter is to let the husband/wife know that the company appreciates his/her support for their spouse. Always try to include a small gift certificate or small gift with a letter like this. Doing so will get you far better results. If you get a gift certificate from a nice store in your area, the spouse can get something out of the ordinary for themselves. A gift certificate in the amount of one or two hours of wages should pay you back many times over. For unmarried workers, change the wording of the letter slightly and send it to a girlfriend or boyfriend or to the employee.

Negative Reference Letter

This is a response to an outside party's request for information regarding someone who used to work for you. In this letter the employee's problem was alcohol, though it could easily be changed to any other subject. However, you might consider delivering your reference in the form of a phone call since that leaves no permanent record.

Positive Reference Letter

This is a positive and enthusiastic reference for a former employee. In this letter the person was a secretary, though you can easily change it as is required.

Introduction Request Letter

This letter is from a subcontractor to a general contractor requesting plans and specs for bidding a specific project. But, more importantly, it is a introduction and an expression of a desire to get to know each other and, hopefully, to become business partners.

Résumé Response

Writing a letter to turn down someone who is looking for employment is never a pleasant exercise. Having a pre-written letter such as this in your word processor's files should be helpful to you. This letter explains that you gave the résumé or application your full attention, and that you are sorry that you have no appropriate positions open at the present time. It also expresses your best wishes for the person finding a good position soon. To this letter, you may wish to add other sections such as one saying that you will keep the résumé in your files or another mentioning another company that may be interested in the person's services. Note that this letter refers to the prospective employee's résumé. Since the use of a résumé has become so common, the letter is written as you see it. If, however, you use job applications, simply substitute the word "application" where "résumé" is shown.

Retirement Letter

This is a letter to a retiring employee. Assuming that the retiring person has worked for you for many years, he or she is not only an employee, but a friend as well. This letter expresses your friendship and good wishes for the employee's future years. It also invites him or her to stop by the office frequently.

Commendation Letter

This is a brief letter of commendation for a salesman who has done a fine job and contributed substantially to the company's success. It can be modified and made appropriate for almost any type of employee.

Company Planning Meeting Topics

- What are our assets?
- What are our liabilities?
- What does the market look like for next year?
- What are the long-term forecasts for the markets we are in?
- Should we start repositioning ourselves into different markets?
- Who are our chief competitors?
- What are our competitor's strengths and weaknesses?
- What are our strengths and weaknesses?
- Can we make a profit operating the way we are now?
- How much profit did we make last year?
- Did our overhead come in within budget?
- Was last year's budget correct?
- Were we over or under?
- Were our labor rates what we expected?
- What will they be next year?
- Was our labor productivity as good as we expected?
- How can we improve our productivity?
- Should we look into a different market? (Contract-bid work, design/build, maintenance, service work, specialty work, etc.)
- What will it take to get into these markets?
- Are there any profitable new markets opening up?
- Who are our key employees?
- Are they happy working with us?
- What can we do to assure that they will stay with us?
- Who are our best customers?
- Are they happy with the service they receive from us?
- What are their most common complaints?
- How can we correct them?
- Do we need to find some new customers?
- How can we promote our services?
- What is our most profitable type of work?
- Can we get more of these types of jobs?

 File Type: Word | **File Name:** 01-Company-Meeting.doc

Company Planning Meeting Topics *(cont.)*

- What are our least profitable types of jobs?

- Do we really need them?

- What were our chief problems last year?

- Why did they happen?

- How can we prevent them in the future?

- Are our workers' attitudes good?

- What can we do to improve their attitudes?

- Do our office people have a good attitude?

- Are we handling our office work efficiently?

- What improvements can be made?

- Are we giving our company enough supervision?

- Is anyone overloaded with work?

- How can we correct this problem?

- Is there anyone who doesn't have enough to do?

- How can we correct this?

- Will we need more people to implement our plans next year?

- Is our management and supervision system working well?

- Where can improvements be made?

- Are there any personality conflicts in the company?

- How can they be corrected?

- Have we had problems with late pay or bad debts?

- Should we change our credit and collections policy?

- Are there any new laws that we need to consider? (Tax, laws, employee regulations, lien laws, etc.)

- Do we need to change our financial structure?

- Do we need more or less insurance?

- Do we have sufficient credit to operate well?

- What are our long-term plans?

- Should they be updated?

Overhead Calculations

Date: _____

Gross income this year: _____

Gross income last year: _____

EXPENSE	LAST YR	THIS YR	% LAST YR	% THIS YR
Owner's/admin. wages	_____	_____	_____	_____
Other office wages	_____	_____	_____	_____
Payroll taxes	_____	_____	_____	_____
Corporate taxes	_____	_____	_____	_____
State and local taxes	_____	_____	_____	_____
Property taxes	_____	_____	_____	_____
Office and shop rent	_____	_____	_____	_____
Utilities	_____	_____	_____	_____
Office supplies	_____	_____	_____	_____
Telephone	_____	_____	_____	_____
Computer and software	_____	_____	_____	_____
Internet	_____	_____	_____	_____
Bank charges	_____	_____	_____	_____
Advertising	_____	_____	_____	_____
Legal fees	_____	_____	_____	_____
Accounting fees	_____	_____	_____	_____
Licenses and dues	_____	_____	_____	_____
Travel, hotels, and meals	_____	_____	_____	_____
Truck expense	_____	_____	_____	_____
Associations and publications	_____	_____	_____	_____
Meals/entertainment	_____	_____	_____	_____
Bad debts	_____	_____	_____	_____
Tools and equipment	_____	_____	_____	_____
Depreciation	_____	_____	_____	_____
Repairs	_____	_____	_____	_____
_____	_____	_____	_____	_____
_____	_____	_____	_____	_____
_____	_____	_____	_____	_____
_____	_____	_____	_____	_____
_____	_____	_____	_____	_____
_____	_____	_____	_____	_____
_____	_____	_____	_____	_____

Project Cost Data Sheet

Job:_____

Start:_____Complete:_____

Size:_____Floor to Floor:_____

Excavation/Site:_____ $ _____

Site Utilities:_____ $ _____

Site Improvements/Landscaping:_____ $ _____

Paving:_____ $ _____

TOTAL SITE:_____ $ _____

Foundation/Slabs:_____ $ _____

Structure:_____ $ _____

Exterior:	Type _____	_____ Sq. Ft.	$ _____
Roof:	Type _____	_____ Sq. Ft.	$ _____
Drywall:	Exterior _____	_____ Sq. Ft.	$ _____
	Interior _____	_____ Sq. Ft.	$ _____
Carpentry:	Rough Finish _____		$ _____
Windows:	Type _____	_____ Sq. Ft.	$ _____
Finishes:	Act _____	_____ Sq. Ft.	$ _____
	Paint _____	_____ Sq. Ft.	$ _____
	VCT _____	_____ Sq. Ft.	$ _____
	Ceramic_____	_____ Sq. Ft.	$ _____
	Carpet _____	_____ Sq. Ft.	$ _____

Elevators:_____ $ _____

Plumbing:_____ $ _____

HVAC:_____ $ _____

_____ $ _____

Sprinkler:_____ $ _____

Electric:_____ $ _____

_____ $ _____

Other Costs:_____ $ _____

General Conditions:_____ $ _____

TOTAL COST:_____ $ _____

OVERHEAD AND PROFIT:_____ $ _____

CONTRACT AMOUNT:_____ $ _____

File Type: Word | **File Name:** 03-Project-Cost-Data.doc

Conditions of Employment

An agreement between _____ (company name), (employer), and _____ (name), employee.

As of this __th day of _____, 20__, the employer and employee agree on the following terms of employment:

1. The attached list of responsibilities outlines the work that will be required of the employee. In addition, other reasonable responsibilities may be required of the employee, who shall be clearly notified of said responsibilities at such a time.

2. The employee's failure to perform the functions mentioned in #1 will be grounds for warning or termination of employment, at the discretion of the employer.

3. Any on the job use of alcohol or any illegal drug by the employee will be grounds for immediate termination of employment.

4. The employee must be sober and alert when arriving at work. Intoxicated employees will not be allowed to work. Such lost time of work will not be recompensed to the employee; and, if occurring more than once, will be grounds for immediate dismissal.

5. Any reckless driving, or other irresponsible and/or dangerous activities will be grounds for warning or dismissal, at the discretion of the employer.

6. No form of personal or sexual harassment will be allowed. Any such activities will be grounds for immediate termination of employment. Any employee convicted of such offenses outside the work place will be immediately terminated as well, since they pose a legitimate threat to the company's other employees.

7. In recompense for the employee's services, he/she will be paid at the following rate: (State weekly, bi-weekly, or monthly wages.)

8. The employee will also receive the following benefits: (State the fringe benefits that the employee is to receive, such as insurance, company car, etc.)

9. In addition to other reasons mentioned in this agreement, employment may be terminated, if such is in the best interest of the employer for financial or other reasons.

10. If the employee is terminated for reasons unrelated to poor performance of his or her duties, the employee shall be entitled to two weeks severance pay. Such wages will be paid no later than the next general payday following the termination of employment.

11. Any form of personal or sexual abuse that occurs on the job must be reported immediately to an officer of the company. It is the policy of _____ (name of company) to immediately terminate any persons engaging in such activity; and the company shall not be held responsible for such abuse, unless the person committing same has been reported to the company and appropriate actions are not taken. In all other cases, the employee agrees that the person initiating such activity is responsible, holding the employer harmless.

12. The employer reserves the right to change this agreement at any time. In such a case, the employee will be notified of the changes and asked to agree or disagree with such modifications. If the employee agrees, employment will continue. If he or she does not agree, employment may be terminated in accordance with items number 9 and 10 of this agreement.

Terms of agreement accepted this day by:

_____   _____

Employee   Representative of employer

_____   _____

Witness   Witness

File Type: Word | File Name: 04-Conditions-Employment.doc

Hourly Labor Rates

ABC CONTRACTORS

123 Any Street
Anytown, US 00000
555-555-5555

Classification: _____

_____

From: _____

To: _____

	Straight Time	Time and One-half	Double Time
Base Labor Rate:			
Hourly Rate	$_____	$_____	$_____
Vacation Pay	$_____	$_____	$_____
Total Base Labor Rate:	$_____	$_____	$_____

Labor Burden:

Payroll Taxes & Insurance:

		Straight Time	Time and One-half	Double Time
FICA	_____%	$_____	$_____	$_____
Federal Unemployment	_____%	$_____	$_____	$_____
State Unemployment	_____%	$_____	$_____	$_____
Workers' Compensation	_____%	$_____	$_____	$_____
General Liability	_____%	$_____	$_____	$_____

Benefits:

		Straight Time	Time and One-half	Double Time
Medical Insurance	$_____/hr.	$_____	$_____	$_____
Other	$_____/hr.	$_____	$_____	$_____
Total Labor Burden:		$_____	$_____	$_____
Total Hourly Labor Cost:		$_____	$_____	$_____
As a Percentage of Wage:		_____%	_____%	_____%

Project Survey

Job Name:	Date:
Address:	
Contact:	Phone:

Notes:

Employee Evaluation

ABC CONTRACTORS

123 Any Street
Anytown, US 00000
555-555-5555

Date: _____   Start Date: _____

Employee: _____   Last Review: _____

Job Title: _____   Evaluated by: _____

Rating:

1. Unsatisfactory
2. Minimum expectations met
3. Average, expected level
4. Consistently exceeds requirements
5. Exceptional performance

Instructions:

Fill in the rating that best fits the employee's overall performance. The rating must be accompanied by a narrative. The overall rating is the average of all the categories.

Quality of Work:

_____   _____
_____
_____

Quantity of Work:

_____   _____
_____
_____

Attitude/Relationships with Others:

_____   _____
_____
_____

Initiative and Self Reliance:

_____   _____
_____

Other:

_____   _____
_____

Daily Time and Labor Distribution Report

Job No.:   Date:

Name:

Location:

Comments:

Weather:

Temp.:

TRADE	EMPLOYEE NAME	EMPLOYEE NUMBER	LABOR CLASSIFICATION	HOURS	RATE	AMOUNT
				TOTALS		

Employee Data

Date: _____

Employee Name: _____

Address: _____

_____

_____

Social Security No.: _____

Date of Birth: _____

Phone Number: _____

Email: _____

Emergency Contacts: _____

_____

_____

Position: _____ Rate of Pay: _____

Notes: _____

_____

_____

_____

_____

_____

_____

Start Date: _____ Hired by: _____

Emergency Medical Information: _____

_____

_____

_____

Employee Time Sheet

ABC CONTRACTORS

123 Any Street
Anytown, US 00000
555-555-5555

Employee Name: _____

Week Ending: _____   Employee No.: _____

Monday:

Job name	Job #	Work code	Hrs

Hours: _____

Thursday:

Job name	Job #	Work code	Hrs

Hours: _____

Tuesday:

Job name	Job #	Work code	Hrs

Hours: _____

Friday:

Job name	Job #	Work code	Hrs

Hours: _____

Wednesday:

Job name	Job #	Work code	Hrs

Hours: _____

Saturday:

Job name	Job #	Work code	Hrs

Hours: _____

Total hours: _____

Emergency Contacts

(Must be filled out BEFORE beginning work on each site.)

Job:	Phone Contact:

Street Name:

Address:

Name	**Contact**	**Phone/Day**	**Phone/Eve.**
Owner			
Contractor			
Superintendent			
Ambulance/EMS			
Hospital/Clinic			
Police			
Fire			
Gas Co.			
Electric Co.			
Telephone Co.			
Underground Service			

Directions to local hospital/clinic:

TO BE POSTED IN FIELD OFFICE AND COPIES DISTRIBUTED TO ALL PERSONNEL.

File Type: Excel | **File Name:** 11-Emergency-Contacts.xls

Emergency Response to Hazardous Substances

If any substance is found of unknown origin, **LEAVE IT ALONE!** Immediately evacuate the area and contact the nearest hazardous material response team. Do not allow employees on site until declared safe by the response team.

FIRST AID

- Arrangements to provide for prompt medical response in the event of an emergency must be made BEFORE starting the project.

- In areas where severe bleeding, suffocation, or severe electrical shock can occur, a 3 to 4 minute response time is required.

- If medical attention is not available within 4 minutes, then a person trained in first aid must be available on the jobsite at all times.

- An appropriate, weatherproof first aid kit must be on site and must be checked weekly.

- Provisions for an ambulance or other transportation must be made in advance.

- Contact methods must be provided.

- Telephone numbers must be posted where 911 is not available.

The following person has adequate training to render first aid in the event of a medical emergency in areas where emergency response time is in excess of 4 minutes: _____

First aid kits are located at the following locations:

- _____   ▪ _____

Every employee shall be trained in the following emergency procedures:

- Evacuation plan
- Alarm systems
- Shutdown procedures for equipment
- Types of potential emergencies

It is the employer's responsibility to review their job sites, addressing all potential emergency situations.

POLICY STATEMENT

(COMPANY NAME) will preserve the safety and health of all employees. We will provide the resources necessary to manage, control, or eliminate safety and health hazards. We will not ignore on-the-job threats to the safety or health of our employees.

All employees are responsible for working safely and productively, as well as recognizing and informing the company of hazards in their work areas.

Employees are also responsible for following safe work practices, including the use of Personal Protective Equipment (PPE) where necessary.

_____

(Company Name) President

 File Type: Word | **File Name:** 12-Emergency-Response.doc

Accident Report

Name of Injured: _____

Social Security No.: _____

Home address of injured: _____

_____

_____

Employer: _____

Age: _____   Male   Female   (circle)

Occupation: _____   How long? _____

Date of injury: _____   Time of accident: _____ A.M. P.M.

Place of accident: _____

_____

_____

Type and nature of injury:

_____

_____

_____

What was employee doing at time of injury?

_____

_____

_____

Where and how did the accident occur?

_____

_____

_____

Specify tool, equipment, substance, or object that directly injured employee:

_____

_____

_____

Accident Report (cont.)

Was medical treatment sought?

Yes _____   No _____

Where and by whom?

_____

_____

_____

Was employee able to work after injury? _____

If no, for how long was he absent from job? _____

Names and addresses of witnesses:

_____

_____

_____

This report filed by: _____

Date: _____

Corrective action taken:

_____

_____

_____

Unsafe conditions or acts contributing to accident:

_____

_____

_____

Explain specifically the corrective actions taken:

_____

_____

_____

Safety Plan

Goal

The primary goal of (COMPANY NAME) is to operate a profitable business by serving its customers.

A primary element of reaching this goal is to keep our employees free from injuries, illness, or harm on the job. We will achieve this, in part, by delegating responsibility and accountability to all involved in this company's operation.

Our safety goals are the following:

- Minimize or eliminate all injuries and accidents
- Minimize loss of property and equipment
- Eliminate all OSHA fines
- Reduce workers' compensation costs
- Reduce operating costs

Specific Steps

We will achieve our safety goals by:

- Appointing well-trained people as our safety coordinators.
- Providing all necessary safety training, especially to safety coordinators.
- Establishing company safety goals.
- Securing honest safety feedback and information from our jobsites. All employees must be able to keep us informed as to safety and health threats.
- Adapting company actions as required to meet safety objectives.
- Developing and implementing a written safety and health program.
- Holding all employees accountable for performance of safety responsibilities.
- Reviewing the safety and health program annually and revising or updating it as required.

Safety Manager

A safety manager shall be appointed to review all safety issues with both field and office personnel.

The safety manager shall:

- Gather all relevant safety information from all sources.
- Discuss safety policies and procedures with all involved.
- Make recommendations for improvements.
- Review accident investigation reports on all accidents and near misses.
- Identify unsafe conditions and work practices and enforce corrections.

The safety manager is authorized to shut down projects, without consultation, upon discovery of any serious threat to our employees.

Safety Plan *(cont.)*

Notice to all Employees

_____ has been designated as our Safety Manager. His/her cell phone and office phone numbers are:

Office: (_____) _____-_____

Cell: (_____) _____-_____

It is the duty of the Safety Manager to assist all of you in keeping our jobs safe.

Please contact _____ immediately regarding any on-the-job health or safety issues. This is one of _____'s primary areas of responsibility.

Our Safety Manager is also responsible for:

- Introducing our safety program to new employees.
- Following up suggestions made by employees. Documenting suggestions and responses.
- Assisting personnel in the execution of safety policies.
- Conducting safety inspections on a periodic basis.
- Addressing all hazards or potential hazards as needed.
- Preparing monthly accident reports and investigations.
- Maintaining an adequate stock of first aid supplies and other safety equipment to insure availability.
- Making sure there are an adequate number of employees that are certified in first aid.
- Staying current with OSHA regulations and local safety mandates.

Supervisor/Foreman

It is the responsibility of our supervisors and foremen to establish an operating atmosphere that insures that safety and health are managed carefully.

Supervisors and foremen are required to do the following:

- Regularly emphasize that accident and health-hazard-exposure prevention are a condition of employment.
- Identify operational oversights that could contribute to accidents.
- Participate in safety- and health-related activities, including attending safety meetings, reviewing the facility, and correcting employee behavior that can result in accidents and injuries.
- Spend time with each person hired to explain the safety policies and the hazards of his/her particular work.
- Make sure that a "Competent Person" is present as required.
- Do not allow safety to be sacrificed for expediency, nor allow workers to do so.
- Enforce safety rules consistently. Follow the company's discipline and enforcement procedures.
- Conduct daily job-site safety inspections and correct safety violations.

File Type: Word | **File Name:** 14-Safety-Plan.doc

Safety Plan *(cont.)*

Employee Responsibilities

It is the duty of each and every employee to know the safety rules, and to conduct his or her work in compliance with these rules. Disregard of the safety and health rules shall be grounds for disciplinary action up to and including termination. It is also the duty of each employee to make full use of the safeguards provided for their protection. Every employee will receive safety instructions when hired and will receive a copy of the Company Safety and Health Program.

Employees are responsible to:

- Read, understand, and follow safety and health rules and procedures.
- Wear Personal Protective Equipment (PPE) at all times when working in areas where there is a danger of injury.
- Wear suitable work clothes as determined by the supervisor or foreman.
- Perform all tasks safely, as directed by a supervisor or foreman.
- Report all injuries to a supervisor or foreman, and seek treatment promptly.
- Know the location of first aid, fire fighting equipment, and other safety devices.
- Attend all required safety and health meetings.
- Not perform potentially hazardous tasks or use any hazardous material unless properly trained to do so. Follow all safety procedures.
- Stop and ask questions if in doubt about the safety of any operation.

Discipline and Enforcement

(COMPANY NAME) maintains standards of employee conduct and supervisory practices which support and promote effective and safe business operations. These supervisory practices include administering corrective action when employee safety performance or conduct jeopardizes this goal. This policy sets forth general guidelines for a corrective action process aimed to document and correct undesirable employee behavior. Major elements of this policy include:

A. Constructive criticism/instruction by the employee's supervisor/foreman to educate and inform employees of appropriate safety performance and behavior.

B. Correcting employee's negative behavior to the extent required.

C. Informing the employee that continued violation of company safety policies can result in termination.

D. Written documentation of disciplinary warnings and corrective action taken.

Depending on the facts and circumstances involved with each situation, the company may choose any corrective action including immediate termination. However, in most circumstances the following steps will be followed:

1. **VERBAL WARNING** informally documented by supervisor, foreman, or safety manager for minor infractions of company safety rules. Supervisor, foreman, or safety manager must inform the employee what safety rule or policy was violated and how to correct the problem.

Safety Plan *(cont.)*

2. **WRITTEN WARNING** documented in employee's file. Repeated minor infractions or a more substantial safety infraction requires issuance of a written warning. The employee should acknowledge the warning by signing the document before it is placed in personnel file.

3. **SUSPENSION** for three (3) working days. If employee fails to appropriately respond or if management determines the infraction was sufficiently grievous.

4. **TERMINATION** for repeated or serious safety infractions.

Control of Hazards

Where feasible, workplace hazards are prevented by effective design of the job site or job. Where it is not feasible to eliminate such hazards, they must be controlled to prevent unsafe and unhealthy exposure. Once a potential hazard is recognized, the elimination or control must be done in a timely manner. These procedures include measures such as the following:

- Maintaining all extension cords and equipment.
- Ensuring all guards and safety devices are working.
- Periodically inspecting the worksite for safety hazards.
- Establishing a medical program that provides applicable first aid to the site as well as nearby physician and emergency phone numbers.
- Addressing any and all safety hazards with employees.

Fire Prevention

Fire prevention is an important part of protecting employees and company assets. Fire hazards must be controlled to prevent unsafe conditions. Once a potential hazard is recognized, it must be eliminated or controlled in a timely manner. The following fire prevention requirements must be met for each site:

- One conspicuously located 2A fire extinguisher (or equivalent) for every floor.
- One 2A conspicuously located fire extinguisher (or equivalent) for every 3000 sq. ft.
- A conspicuously located 10B fire extinguisher for everywhere more than 5 gallons of flammable liquids or gas are stored.
- Generators and internal combustion engines located away from combustible materials.
- Site free from accumulation of combustible materials or weeds.
- No obstructions or combustible materials piled in the exits.
- No more than 25 gallons of combustible liquids stored on site.
- No Liquid Propane Gas containers stored in any buildings or enclosed spaces.
- Fire extinguishers in the immediate vicinity where welding, cutting, or heating is being done.

Safety Plan *(cont.)*

Training and Education

Training is an essential component of an effective safety and health program addressing the responsibilities of both management and employees at the site. Training is most effective when incorporated into other education on performance requirements and job practices. Training programs should be provided as follows:

- Initially when a safety and health plan is developed.
- For all new employees before beginning work.
- When new equipment, materials, or processes are introduced.
- When procedures have been updated or revised.
- When experiences/operations show that employee performance must be improved.
- At least annually.

Besides the standard training, employees should also be trained in the recognition of hazards; to be able to look at an operation and identify unsafe acts and conditions. A list of typical hazards employees should be able to recognize may include:

- **Fall Hazards**—Falls from floors, roofs and roof openings, ladders (straight and step), scaffolds, wall openings, hazards from tripping, trenches, steel erection, stairs, chairs, etc.
- **Electrical Hazards**—Appliances, damaged cords, outlets, overloads, overhead high voltage, extension cords, portable tools (broken casing or damaged wiring), grounding, metal boxes, switches, ground fault interrupters (GFI), etc.
- **Housekeeping Issues**—Exits, walkways, floors, trash, storage of materials (hazardous and non-hazardous), protruding nails, etc.
- **Fire Hazards**—Oily-dirty rags, combustibles, fuel gas cylinders, exits (blocked), trips/slips, stairs, uneven flooring, electrical cords, icy walkways, etc.
- **Health Hazards**—Silicosis, asbestos, loss of hearing, eye injury due to flying objects

Employees trained in the recognition and reporting of hazards and supervisors/foremen trained in the correction of hazards will substantially reduce the likelihood of a serious injury.

Recordkeeping and OSHA Log Review

In the event of a fatality (death on the job) or catastrophe (accident resulting in hospitalization of three or more workers) contact (Safety Manager). His/her office and cell-phone numbers are:

Office: (_____) _____–_____

Cell: (_____) _____–_____

The Safety Manager will in turn report it to the OSHA Regional Office within 8 hours after the occurrence.

If an injury or accident should ever occur, you are to report it to your supervisor or foreman as soon as possible. A log entry and summary report shall be maintained for every recordable injury and illness. The entry should be done within 7 days after the injury or illness has occurred. An appropriate form shall be used for the recording. A recordable injury or illness is defined as an injury resulting in loss of consciousness, days away from work, days of restricted work, or medical treatment beyond first aid.

File Type: Word | File Name: 14-Safety-Plan.doc

Safety Plan *(cont.)*

First Aid includes:

- Tetanus shots
- Band-aids or butterfly bandages
- Cleaning, flushing, or soaking wounds
- Ace bandages and wraps
- Non-prescription drugs at non-prescription strength (Aspirin, Tylenol, Etc.)
- Drilling fingernails/toenails
- Eye patches, eye flushing, and foreign body removal from eye with Q-tips
- Finger guards
- Hot or cold packs
- Drinking fluids for heat stress

An annual summary of recordable injuries and illnesses must be posted at a conspicuous location in the workplace and contain the following information: Calendar year, company name, establishment name, establishment address, certifying signature, title, and date. If no injury or illness occurred in the year, zeros must be entered on the total line.

The OSHA logs should be evaluated by the employer to determine trends or patterns in injuries in order to appropriately address hazards and implement prevention strategies.

Accident Investigation

Supervisors/Foreman

- Provide first aid, call for emergency medical care if required.
- If further medical treatment is required, arrange to have an employer representative accompany the injured employee to the medical facility.
- Secure area, equipment, and personnel from injury and further damage.
- Contact Safety Manager.

Safety Manager

- Investigate the incident (injury). Gather facts, employee, and witness statements; take pictures and physical measurements of incident site and equipment involved.
- Complete an incident investigation report form and the necessary workers' compensation paperwork within 24 hours whenever possible.
- Insure that corrective action to prevent a recurrence is taken.
- Discuss the incident, if appropriate, in safety and other employee meetings with the intent to prevent a recurrence.
- Discuss incident with other supervisors, foremen and other management personnel.
- If the injury warrants time away from work, be sure that the absence is authorized by a physician and that you maintain contact with your employee while he/she remains off work.

Safety Plan *(cont.)*

- Monitor status of employee(s) off work, maintain contact with employee and encourage return to work even if restrictions are imposed by their physician.

- When injured employee(s) return to work, they should not be allowed to return to work without "return to work" release forms from the physician. Review the release carefully and insure that you can accommodate the restrictions, and that the employee follows the restrictions indicated by the physician.

Safety Rules and Procedures

- No employee is expected to undertake a job until that person has received adequate training.

- All employees shall be trained on every potential hazard that they could be exposed to and how to protect themselves.

- Employees are not required to work under conditions which are unsanitary, dangerous, or hazardous to their health.

- Only qualified trained personnel are permitted to operate machinery or equipment.

- All injuries must be reported to your supervision/foreman.

- Manufacturer's specifications/limitations/instructions shall be followed.

- Particular attention should be given to new employees and to employees moving to new jobs or doing non-routine tasks.

- Emergency numbers shall be posted and reviewed with employees.

- Each employee working in an excavation or trench shall be protected from cave-ins by an adequate protective system.

- Employees working in areas where there is a danger of head injury, excessive noise exposure, or potential eye and face injury shall be protected by Personal Protection Equipment (PPE).

- All hand and power tools and similar equipment, whether furnished by the employer or the employee, shall be maintained in a safe condition.

- All materials stored in tiers shall be stacked, racked, blocked, interlocked, or otherwise secured to prevent sliding, falling, or collapse.

- The employer shall insure that electrical equipment is free from recognized hazards that are likely to cause death or serious physical harm to employees.

- All scaffolding shall be erected in accordance with the CFR 1926.451 subpart L. standard. Guardrails for fall protection and ladders for safe access shall be used.

- Places of employment shall be kept clean, the floor of every workroom shall be maintained, so far as practicable, in a dry condition; standing water shall be removed. Where wet processes are used, drainage shall be maintained and false floors, platforms, mats, or other dry standing places or appropriate waterproof footgear shall be provided.

Safety Plan *(cont.)*

- To facilitate cleaning, every floor, working place, and passageway shall be kept free from protruding nails, splinters, loose boards, and holes and openings.

- All floor openings, open sided floor, and wall openings shall be guarded by a standard railings and toe boards or cover.

- The employer shall comply with the manufacturer's specifications and limitations applicable to the operation of any and all cranes and derricks.

- All equipment left unattended at night, adjacent to a highway in normal use, or adjacent to construction areas where work is in progress, shall have appropriate lights or reflectors, or barricades equipped with appropriate lights or reflectors, to identify the location of the equipment.

- No construction loads shall be placed on a concrete structure or portion of a concrete structure unless the employer determines, based on information received from a person who is qualified in structural design, that the structure or portion of the structure is capable of supporting the loads.

- A stairway or ladder shall be provided at all personnel points of access where there is a break in elevation of 19 inches or more, and no ramp, runway, sloped embankment, or personnel hoist is provided.

Employee Emergency Action Plan

Fires and Other Emergencies

The following emergency action plan is appropriate only for small construction sites, larger sites should have a much more detailed plan.

1. **Emergency escape procedures:** Immediately leave the building through the closest practical exit. Meet up at the foremen's truck.

2. **Critical plant operations:** Shut off the generator on your way out if possible, otherwise evacuate the building.

3. **Accounting for Employees:** Foreman/Supervisor is to account for all employees after emergency evacuation has been completed and is to assign duties as necessary.

4. **Means of reporting fires and other emergencies:** Dial 911 on the cell phone, report the location of the emergency, and provide directions to the responders.

5. **Further Information:** Contact the Safety Coordinator for further information or explanation of duties under the plan.

TRAINING: Before implementing the emergency action plan, a sufficient number of persons to assist in the safe and orderly emergency evacuation of employees will be designated and trained. The plan will be reviewed with each employee covered by the plan at the following times:

1. Initially when the plan is developed or upon initial assignment.

2. Whenever the employee's responsibilities or designated actions under the plan change.

3. Whenever the plan is changed.

The plan will be kept at the worksite and made available for employee review. For those employers with 10 or fewer employees the emergency action plan may be communicated orally to employees and the employer need not maintain a written plan.

Policies and Procedures Acknowledgements

I have read and understand the attached company policies and procedures and agree to abide by them. I have also had the duties of the position which I have accepted explained to me, and I understand the requirements of the position. I understand that any violation of the above policies is reason for disciplinary action up to and including termination.

_____   _____

Employee Signature   Date

Truck Inspection Sheet

ABC CONTRACTORS

123 Any Street
Anytown, US 00000
555-555-5555

Date: _____

Type of Vehicle: _____   License No.: _____

Make of Vehicle: _____   Mileage: _____

CONDITION				
Body	**OK**	**Marginal**	**Repair**	**Remarks**
Paint				
Lights				
Turn Signals				
Windshield				
Chassis	**OK**	**Marginal**	**Repair**	**Remarks**
Clutch				
Fire Extinguisher				
Flashlight/Flares				
Foot Brake				
Hand Brake				
Seat Belts				
Shocks				
Spare Tire				
Tires				

Truck Inspection Sheet (cont.)

CONDITION				
Motor	**OK**	**Marginal**	**Repair**	**Remarks**
Air Cleaner				
Ammeter				
Battery				
Engine Operation				
Fuel Gauge				
Horn				
Hoses				
Oil Filter				
Oil Pressure				
Temperature Gauge				

Other

Accident instructions in glove box	[ ] Yes	[ ] No
Evidence of insurance	[ ] Yes	[ ] No
Log up to date	[ ] Yes	[ ] No
Registration	[ ] Yes	[ ] No
Spare keys	[ ] Yes	[ ] No

Submitted By: _____

Travel Expense Report

ABC CONTRACTORS

123 Any Street
Anytown, US 00000
555-555-5555

Date of Request: _____

Employee: _____

Travel Period, From: _____

Travel Period, To: _____

Date:	Sun	Mon	Tue	Wed	Thur	Fri	Sat	TOTAL	Details:
Starting Mileage									
Ending Mileage									
Mileage/Day									
Reimburse/Mile									
Air/Ground Fare									
Auto Rental									
Parking									
Tolls									
Lodging									
Telephone									
Meals									

Signature: _____   Date: _____   Approved: _____

Equipment Ledger

ABC CONTRACTORS

123 Any Street
Anytown, US 00000
555-555-5555

Date: _____   Date Acquired: _____

Item #: _____   Description: _____

Initial Cost: _____   Estimated Life: _____

Estimated Use Per Year: _____   Est. Salvage Value: _____

Avg. Annual Investment: _____   Annual Deprec.: _____

Year	Item	Operating hours	Cumulative hours	Cost per hour
1	Depreciation			
	Interest, taxes, insurance			
	Fuel, oil, lube			
	Repairs, parts			
	Tires			
	Operating labor			
	TOTAL			
2	Depreciation			
	Interest, taxes, insurance			
	Fuel, oil, lube			
	Repairs, parts			
	Tires			
	Operating labor			
	TOTAL			
3	Depreciation			
	Interest, taxes, insurance			
	Fuel, oil, lube			
	Repairs, parts			
	Tires			
	Operating labor			
	TOTAL			

File Type: Excel | **File Name:** 18-Equipment-Ledger.xls

Employee Driving Record

ABC CONTRACTORS

123 Any Street
Anytown, US 00000
555-555-5555

Date: _____   Phone: _____

Employee: _____   Driver's Lic. No.: _____

Address: _____   State: _____

Please answer the following questions:

1. Has your license ever been suspended or revoked?   [ ] Yes [ ] No

   If yes, please explain: _____

   _____

2. Have you ever been cited for driving under the influence?   [ ] Yes [ ] No

   If yes, please explain: _____

   _____

3. Have you received any moving violations within the past 3 years?   [ ] Yes [ ] No

   If yes, please explain: _____

   _____

4. Have you been involved in any accidents within the past 3 years?   [ ] Yes [ ] No

   If yes, please explain: _____

   _____

I give my permission for you to obtain a verification of my driving record.

I certify that the above information is true and correct.

Signed: _____   Date: _____

File Type: Excel | **File Name:** 19-Employee-Driving-Record.xls

Cash Paid Out

ABC CONTRACTORS

123 Any Street
Anytown, US 00000
555-555-5555

Date: _____   Date Paid: _____

Project #: _____   Check #: _____

Project Name: _____   Amount: _____

Paid To: _____

_____

_____

For: _____

_____

_____

_____

_____

_____

_____

_____

_____

_____

Safety Agreement

ABC CONTRACTORS

123 Any Street
Anytown, US 00000
555-555-5555

Date: _____

Employee: _____

Social Security No.: _____

The undersigned, _____, hereby acknowledges and agrees that:

1. I have been shown the location of the first-aid kit, emergency telephone numbers, and fire extinguisher.

2. I have received instruction in the use of the fire extinguisher, safety goggles, hard hat, and other equipment applicable to my trade.

3. I have been issued a hard hat and will wear it at all times when on the job site.

4. I have received a copy of the Job Site Safety Program; I have read it, I understand it, and I will comply with each and every provision.

5. In case of injury, I will report it to my supervisor.

6. I will report any unsafe conditions to my supervisor.

Signed: _____   Date: _____

Employee Warning Letter #1

ABC CONTRACTORS

123 Any Street
Anytown, US 00000
555-555-5555

(Date)

(Name)

(Address)

(City, State, Zip)

Dear _____,

It is always difficult to reprimand an employee whose history with us has been one of mutual satisfaction for both employer and employee. During the time you've spent here, your performance of your duties has been excellent; however, we are very concerned about your increasing frequency of absences, which are affecting not only your work but also our proficiency as a company. We understand you have had health problems and some personal problems. However, one of the requirements for employment with us is regular attendance; when we schedule an employee to work a specific time, we must assume that employee will be present to perform his/her duties.

Though we have attempted to resolve this issue by various methods and through various channels, we remain willing to assist if you can suggest some reasonable alternative to this problem. If you have any suggestions, please contact your supervisor immediately; otherwise, we have no choice but to terminate your employment.

Sincerely,

_____

(Name)

(Title)

Employee Warning Letter #2

ABC CONTRACTORS

123 Any Street
Anytown, US 00000
555-555-5555

(Date)

To: All employees

Re: Substance abuse

Recently it was brought to our attention that some of our employees may be drinking or using illegal drugs during working hours. This memo is to serve as an official notice that such activity will not be tolerated. Not only do we expect our employees to work during working hours, but company policy prohibits the use of any behavior-altering drug on company property, even if the employee is technically on break. Company premises are to remain drug-free. Any employee discovered engaging in such practices will be terminated immediately, without recourse.

Sincerely,

_____

(Name)

(Title)

Termination Letter #1

ABC CONTRACTORS

123 Any Street
Anytown, US 00000
555-555-5555

(Date)

(Name)
(Address)
(City, State, Zip)

Dear _____,

After a long and thoughtful consideration of our company's present financial situation, we concluded that we must eliminate several positions. Regretfully, your position is one of those to be eliminated.

On a more optimistic note, we expect conditions to improve in the near future, and when they do, we will consider you for reemployment if you are interested. Until then, we wish you every success in finding new employment and sincerely thank you for the work you have done for us.

Sincerely,

_____

(Name)
(Title)

Termination Letter #2

ABC CONTRACTORS

123 Any Street
Anytown, US 00000
555-555-5555

(Date)

(Name)
(Address)
(City, State, Zip)

Dear _____ ,

Even though you have been warned numerous times about _____ (identify behavior), your behavior remains unchanged. You leave us no other choice; effective today _____ (date), your employment with _____ (company name) is terminated. We will mail your final paycheck after our next regularly scheduled pay day.

Sincerely,

_____

(Name)
(Title)

Reporting Accidents

ABC CONTRACTORS

123 Any Street
Anytown, US 00000
555-555-5555

(Date)

To: All employees

Re: Reporting of accidents

Dear Employee,

Please remember the following: All accidents on the job must be reported immediately. If this is not done, it could jeopardize any possible worker's compensation.

All job foremen should have the appropriate forms for accident reports. However, if anyone should run out of forms, _____ (name), in the front office, can get you as many extras as you may require.

Again, remember that filling out these forms immediately after any injury is critical.

Sincerely,

_____

(Name)

(Title)

File Type: Word | **File Name:** 26-Reporting-Accidents.doc

Injury Letter

ABC CONTRACTORS

123 Any Street
Anytown, US 00000
555-555-5555

(Date)

(Name)
(Address)
(City, State, Zip)

Dear _____,

I am so sorry to hear about your (accident, injury). I hope that you are on the road to a quick recovery.

Since you will be confined for a few ____ (time period), I think you will enjoy this _____ (gift) for your hobby of _____ (hobby).

Sincerely,

_____

(Name)
(Title)

Insurance Change Notice

ABC CONTRACTORS

123 Any Street
Anytown, US 00000
555-555-5555

(Date)

To: All employees

Re: Insurance requirements

Recently we were informed by our insurance company that certain policy changes must be made or we will lose insurance coverage. Since losing our coverage is not an option, and since we do not have adequate time to shop for new coverage, we are making the adjustments as requested. In an effort to keep you informed, we have compiled a brief but thorough list of the changes being implemented. They are as follows:

(provide details)

Understandably, these changes may prove difficult for some, but they are mandatory. Please be assured that we remain committed to you, and will continue to do what we must to ensure adequate coverage for our employees.

Sincerely,

_____

(Name)

(Title)

Letter of Appreciation

ABC CONTRACTORS

123 Any Street
Anytown, US 00000
555-555-5555

(Date)

(Name)

(Address)

(City, State, Zip)

Dear _____,

I want to express my personal thanks to you for your support of your husband _____ (or other person) these last few weeks. As you know, we've been working overtime to complete the _____ (name of project) in _____ (location). I know this has caused _____ (husband's name) to get home late from work, which must have taken away from your family time.

Because of this, we are sending a small gift certificate to the spouses of all the men who have been working on this project. Buy yourself something nice, you deserve it.

Again, thank you for supporting _____ (name) during this very busy time.

With appreciation,

_____

(Name)

(Title)

Negative Reference

ABC CONTRACTORS

123 Any Street
Anytown, US 00000
555-555-5555

(Date)

(Name)

(Address)

(City, State, Zip)

Dear _____,

In response to the Reference Letter Request for _____, I can verify that _____
was employed by our organization from _____, ____ until _____, ____.

As a/an [insert title], his/her primary duties included: _____
_____.

This letter should serve to neither confirm nor deny that _____'s performance warrants a
positive reference by me or my company. It is my recommendation that you solicit additional requests
from the references provided by _____.

Sincerely,

_____

(Name)

(Title)

Positive Reference

ABC CONTRACTORS

123 Any Street
Anytown, US 00000
555-555-5555

(Date)

(Name)

(Address)

(City, State, Zip)

To Whom It May Concern:

I have had the pleasure of working with _____ over the past _____ and highly recommend [him/her] to any prospective employer. [His/Her] _____ and _____ skills are truly impressive. The contributions [he/she] has made to our team have improved our _____ program/division beyond my expectations.

Over the years, as a manager, I have come to value employees who achieve and work well without close supervision. _____ excels in this area. [He/She] listens closely, understands directions, respects deadlines, can handle multiple projects and has proven on numerous occasions that [he/she] does not need to be closely monitored. [He/She] has a patient and calm personality that allows [him/her] to work well with our team.

_____ is an extraordinary person and is respected among [his/her] coworkers. I have witnessed, first hand, [his/her] skills and assets. I firmly believe [he/she] will go far and has a lot to offer any organization.

Sincerely,

_____

(Name)

(Title)

Introduction Request

ABC CONTRACTORS

123 Any Street
Anytown, US 00000
555-555-5555

(Date)

(Name)

(Address)

(City, State, Zip)

Dear _____,

For quite some time now we have noticed how often your firm is involved in some of the most prestigious projects in our area. Although we have never worked together, our company has a long and solid work history in this area and I think, like you, we have earned the excellent reputation associated with our name because we focus on quality.

I am writing to ask that you add us to your bidding list for _____ (specify) work. We will furnish whatever credentials necessary including references, biographies of key employees, a history/overview of our company, etc., in order to establish a connection.

Because our company has been in business for _____ (length of time) years, and we are professionals in our field, we believe we have something to offer your company. For that reason, I would like to establish a working relationship, which I'm convinced will prove to be a positive thing for both of us. Perhaps we might meet for lunch one day to get acquainted. Thank you.

Sincerely,

_____

(Name)

(Title)

File Type: Word | **File Name:** 32-Request-Intro.doc

Résumé Response

ABC CONTRACTORS

123 Any Street
Anytown, US 00000
555-555-5555

(Date)

(Name)

(Address)

(City, State, Zip)

Dear _____,

Your interest in _____ (company) is genuinely appreciated. I read your resume with interest.

However, after evaluating the resume with our personnel and marketing people, I regret that we do not have a position open at this time that fits your abilities.

We thank you for your interest in _____ (company) and I hope that you will find a good position soon.

Sincerely,

_____

(Name)

(Title)

Retirement Letter

ABC CONTRACTORS

123 Any Street
Anytown, US 00000
555-555-5555

(Date)

(Name)

(Address)

(City, State, Zip)

Dear _____,

Congratulations and best wishes for your retirement. We thank you for your _____ years of productive service.

Many of your ideas have been instituted within our company. We will surely miss you but know that you will enjoy your leisure. Come see us often; we think of you as part of our family!

Sincerely,

_____

(Name)

(Title)

Commendation Letter

ABC CONTRACTORS

123 Any Street
Anytown, US 00000
555-555-5555

(Date)

(Name)
(Address)
(City, State, Zip)

Dear _____,

Congratulations for doing such a fine job of sales. _____ (company) now has an annual sales volume of $_____.

We know that your efforts in securing the _____ (name) contract has definitely helped us continue our success.

Sincerely,

_____
(Name)
(Title)

CHAPTER TWO

Forms and Letters for Your Customers

This chapter contains useful forms, contracts, and letters based on your business correspondence with your customers.

FORMS

Company Evaluation

This is a feedback form to be sent to your customers. This letter asks, "How did we do?" and is useful for improving your operations and as a sales tool. Don't worry if many customers don't return them. The fact that you sent the form says good things about you.

Invoice

This is a basic invoice form for billing the job.

Detailed Invoice

This is a basic, detailed invoice form for a large, ongoing project.

Warranty

This is a formal warranty of specific equipment from a contractor to a customer. It is a legally binding document. Provision is made at the bottom for it to be notarized.

LETTERS

Sales Letter

This is an introductory sales letter for prospective new customers. In it you offer to spend time and effort proving your worth to the client. You will notice that this letter is written primarily for industrial clients. You can, of course, modify it for other types of clients as well. Remember that this letter is selling your service and sincerity to your clients. You must be prepared to provide these services before you send a letter such as this one. This is especially true if you are selling to industrial concerns; they rely deeply on reliability, and if you disappoint them before you have completed several successful projects for them, they will likely never give you another chance at their work. Read this letter carefully before you send it, and make sure that it matches your business philosophy. If it doesn't, modify it before you use it.

Sales Follow-up Letter

This is a simple follow-up letter to a customer you made a proposal to a short time before. Do not neglect simple sales tools such as this one—they keep dialogue open and are more than worth the minimal effort they require.

Customer Thank You Letter

This is a simple end-of-the-year thank-you note; the kind that you might send with a Christmas gift to a good customer. If you use it in this manner, you should include a brief note about the attached gift, such as: "I know that you will enjoy this gift certificate for a night out with your family."

Inclusion Letter

This is a request to be allowed to bid an upcoming project. It is to be sent along with a bid-document deposit.

Inclusion Thank You Letter

A subcontractor's letter should be sent thanking a GC for including their company in bidding. This is another one of the marketing letters that will set your company apart from the crowd. By expressing your appreciation for an opportunity, you go a long way toward assuring that there will be more such opportunities. In addition, the note at the end of the letter requesting feedback can help immensely. Nothing is better than honest feedback from a customer, and it is rare in the construction industry. You should go out of your way to get such information. Think about including notes of this type in other correspondences as well, the information it gets you can be very valuable.

Lost Customer Letter

This is a letter to a former customer. Repeat business is the best kind you can get, so to lose a good customer is a big loss. This letter is written in hope of getting a customer back. At the very minimum, you can use this letter to begin a dialogue with the ex-customer. To show your sincerity on that account, you should enclose a stamped, self-addressed envelope with the letter for the ex-customer's response. You might also ask for an e-mailed response.

Bad Check Letter

This is a letter regarding a bad check written to you by a customer. Always remember, when dealing with payment issues, that the primary concern is not to punish someone, but to get the money.

Collection Letter #1

This is the first of three collection letters. These are written to be used in numerical order.

Collection Letter #2

This is the second of three collection letters. These are written to be used in numerical order.

Collection Letter #3

This is the third of three collection letters. These are written to be used in numerical order.

Guarantee Letter

Guarantees are required for almost all construction projects. The standard time for such a guarantee is one year from the date of acceptance. This date of acceptance and the length of the guarantee's term should be filled in the appropriate blank spaces on this form. The last note in this letter gives you something of an "out" against unreasonable demands for guarantee work. By tying the guarantee to standard trade practices, you have some legal ammunition to use when declining to do unreasonable things. You may also want to simply refer to the document by name and state that a copy will be made available to the customer, should they request it.

Equipment Warranty Letter

This letter regards one of the trickier problems on a construction project, placing equipment into service. The problem here is the warranty coverage. The equipment manufacturer generally warranties equipment for one year from the date it is placed in service (or sometimes, one year from the date of purchase). Your contract agreement, on the other hand, almost assuredly requires you to warrant the equipment for one year from the date of acceptance. Thus, if you place the equipment in service a period of time before the owner issues his or her final acceptance of the project, you are widely exposed during the time gap. Many types of equipment used on construction projects (chillers, switchgear and transformers, pumps, refrigeration and cooling units, HVAC equipment, etc.) are very expensive, and your company could be hurt or destroyed if you had to replace them on your own.

In response to such problems, this letter requests a written acceptance of the systems that are to be placed into service prior to the final acceptance of the entire project. Note that the letter approaches the subject gingerly and does not have a demanding tone. But even though the letter is conciliatory in tone, you should not put this equipment into service without a signed partial acceptance. This may seem like a hard attitude to take, but why should you put yourself and your company at risk? Your contract does not require you to take such risks, so why should you be intimidated into doing risky things? Do them enough times, and you'll eventually be burned.

Late Pay Letter #1

This is a statement of an overdue account. It is a notice of past-due invoices and a request for immediate payment.

Late Pay Letter #2

This is a letter regarding an overdue account. The idea behind this letter is that you wish to salvage the account, and continue working with them in the future. In other words, this is a customer that you like, but one who has fallen far behind in their payments.

Late Pay Letter #3

This is a late payment letter with an enclosure. This is a first or second effort at collecting an overdue account. You will want to use many different types of such letters. This letter uses a postage-paid envelope as an enclosure, which will hopefully encourage prompt payment.

Late Pay Letter #4

This is a friendly reminder notice. It is a mild response to an unpaid balance. Its tone is friendly and almost conciliatory. It works very well as a first reminder.

Payment Received Late Letter

This is an acknowledgement of receiving payment following one or more late notices. These are important to send every time they are applicable.

Qualification Form Letter

This is a general contractor's request for subcontractor to fill out a qualification form. This is a cover letter for a subcontractor qualification form and a solicitation to prospective subcontractors. The general contracting firm sending this letter is looking for certain types of subcontractors and sends out the form to subcontractors of that type. Hopefully, some good subs will fill out the form and begin bidding to the general contractor. This letter could also be slightly modified and used by subcontractors in search of specialty contractors. Since the qualification form that will be included with this letter asks for some very confidential information, a note is included assuring the recipient that all information will be kept strictly confidential.

Rate Increase Letter

In every field, there comes a time when a price increase is necessary. When this happens, you must notify your customers beforehand, and do your best to explain the increase in such a way that maintains their good will toward your company. In such letters, you must be neither too apologetic nor too aloof. Explain why the increase is necessary and that you really don't want to raise your prices. Then, state the new price and the date on which it is effective. As you notice in this letter, you should offer your willingness to discuss such rate increases with good customers. Presumably, you will not be making exceptions to the rate increase. It is helpful to talk to customers to allow you to justify the increase or to work out some other type of billing arrangements.

Referral Thank You Letter

This is a letter of gratitude to someone who recommended your services. It is designed to make this person feel comfortable with recommending you. It does so by first expressing sincere gratitude (everyone likes to feel appreciated), and then by assuring the person that you will take the best possible care of the customer. This final assurance of good treatment is extremely important. You should use letters like this every time you can. Do not use these letters instead of calling on the phone or thanking the party personally; do both. In every trade, people and their opinions are the most important of factors. A quickly prepared note of thanks will set you apart from the crowd in many subtle (yet important) ways.

Refund Letter

This is a conciliatory letter to a customer who complained about an incorrect bill. The details given in this copy of the letter concern a defective product, but could (and should) be changed for whatever situation may face you. When you must refund money or admit to a fault in some other way, do it forthrightly. Send a personal letter with the check or apology, and assure your customer that the overcharging was an accident that will not be repeated.

Company Evaluation

ABC CONTRACTORS

123 Any Street
Anytown, US 00000
555-555-5555

Address: _____   Date: _____

_____   Project Name: _____

1. How did you find out about our company? _____

2. Do you feel your property was respected by our crew and subcontractors?   [   ] Yes [   ] No
   If no, please explain any difficulties you have had:

   _____

   _____

   _____

3. How would you rate our professionalism?
   [   ] excellent   [   ] good   [   ] poor

4. How would you rate our administrative performance?
   [   ] excellent   [   ] good   [   ] poor

5. How do you feel we handled the day-to-day phases of your job?
   [   ] excellent   [   ] good   [   ] poor

6. How would you rate our overall construction performance?
   [   ] excellent   [   ] good   [   ] poor

7. What could we have done to make your project run more smoothly?

   _____

   _____

8. Are you satisfied with your completed project?   [   ] Yes [   ] No
   What, if anything, should have been done differently?

   _____

   _____

9. How well were problems dealt with? _____

10. May we use you as a reference?   [   ] Yes [   ] No

Please include any additional comments.

_____

_____

Thank you.

Invoice

ABC CONTRACTORS

123 Any Street
Anytown, US 00000
555-555-5555

From:	**No.**
To:	Work Performed at:

Date: _____

Work Order No.: _____

Bid No.: _____

Description of Work Performed:

_____

_____

_____

_____

_____

_____

_____

All material is guaranteed to be as specified. The above work was performed in accordance with the drawings and specifications provided for the above. Work was completed in a workmanlike manner for the agreed sum of _____ Dollars ($_____).

This is a ☐ Partial ☐ Full invoice due and payable by:

_____ Month _____ Day _____ Year

in accordance with our ☐ Agreement ☐ Proposal No. _____

Dated _____ Month _____ Day _____ Year

Detailed Invoice

ABC CONTRACTORS
123 Any Street
Anytown, US 00000
555-555-5555

Application #: _____
Application Date: _____
Period: _____

A	B	C	D		E	F	G	H	I
			Work completed						
Item No.	Work Description	Scheduled Value	Previous Application (D + E)	This Period	Presently Stored Materials (not D or E)	Total Completed & Stored to Date (D + E + F)	% (G/C)	Balance to Finish (C – G)	Retainage
TOTALS									

File Type: Excel | **File Name:** 38-Detained-Invoice.xls

Warranty

ABC CONTRACTORS

123 Any Street
Anytown, US 00000
555-555-5555

Date: _____

Project Designation: _____

Address: _____

                   _____

                   _____

The undersigned, _____ (contractor), hereby grants the following warranty to _____ (owner) the which we have installed in the above-named project for _____ (number) years use, beginning _____ (date).

The undersigned hereby agrees to repair or replace, to the satisfaction of the owner, any or all such work that may prove defective in workmanship or materials within that period, ordinary wear and tear excepted, together with any other work which may be damaged or displaced in so doing. If we fail to comply with the above-mentioned conditions within a reasonable time after being notified in writing, we do hereby authorize the owner to have the defects repaired and made good at our expense, and we will pay the costs and charges, including reasonable attorney's fees, upon demand. This warranty covers and includes any special terms, including time periods, specified for this work or materials in the plans, specifications, and contract documents for this project.

Signed,

_____

(Name)

(Title)

State of _____ (state), County of _____ (county).

On the _____ (number) day of _____ (month) in the year _____ (year), before me personally came _____ to me known, who, being by me duly sworn, did depose and say that he resides in _____ (address); that he is the _____ (officer or director title) of the _____ (name of corporation), the corporation described in and which executed the above instrument; and that he signed his name thereto by authority of the board of directors of said corporation.

(Notary Public Seal)

Sales Letter

ABC CONTRACTORS

123 Any Street
Anytown, US 00000
555-555-5555

(Date)

(Name)
(Address)
(City, State, Zip)

Dear _____,

Please accept this letter as an introduction to our company, a full-service _____ (electrical, plumbing, GC, etc.) provider. We have been in business for over _____ (years) and have a long history with satisfied customers.

Currently, we are in the process of expanding our client base and of course, your name is one associated with quality and distinction, qualities we look for in every aspect of our business. Our method of doing business is simple; we seek customers who think of us as a friend, who return for our services and products after years of knowing us. We want to represent a stable, secure source that consistently delivers quality through our products and workmanship. And we want associates who aspire to the same. Judging by your reputation, I think you do.

At your convenience, I would like to stop by your office one day to personally introduce myself to you. Perhaps you could provide something specific regarding your expectations of a _____contractor. I seek no obligation or commitment; meeting personally simply makes for better conversation and a more concise exchange of ideas. I will call in a few days to set up a meeting. Thank you for considering my proposal.

Sincerely,

_____

(Name)

(Title)

Sales Follow-up Letter

ABC CONTRACTORS

123 Any Street
Anytown, US 00000
555-555-5555

(Date)

(Name)
(Address)
(City, State, Zip)

Dear _____,

I am taking the liberty of writing so that I do not interrupt you by phoning. Last week I mailed a proposal, which I believed might result in something positive for both of us, however, I have had no response from you.

At the risk of repeating myself, our company would like to do business with yours; I believe we both have something of value to offer the one another. Please let me know as soon as possible if you are interested in pursuing this alliance.

Cordially,

_____

(Name)

(Title)

Customer Thank You Letter

ABC CONTRACTORS

123 Any Street
Anytown, US 00000
555-555-5555

(Date)

(Name)
(Address)
(City, State, Zip)

Dear _____,

I wanted to write just a quick "thank you" for being one of our customers this year.

We appreciate your business and we hope to continue serving you during the coming year.

Sincerely,

_____

(Name)

(Title)

Inclusion Letter

ABC CONTRACTORS

123 Any Street
Anytown, US 00000
555-555-5555

(Date)

(Name)
(Address)
(City, State, Zip)

Re: _____ (project name or designation)

Dear _____,

We have just learned of the aforementioned project and are interested in bidding it.

As we understand, there is a deposit of $_____.00 required for the bidding documents. We have therefore enclosed that amount, payable to your firm.

Please forward the documents to us as soon as possible, and please put us on your list to receive any and all addenda for the project.

Thank you for your help; and if you should have any questions, please call us at the number shown on this letterhead.

Sincerely,

_____

(Name)
(Title)

Inclusion Thank You Letter

ABC CONTRACTORS

123 Any Street
Anytown, US 00000
555-555-5555

(Date)

(Name)
(Address)
(City, State, Zip)

Re: _____ (project name or designation)

Dear _____ ,

I want to thank you for allowing _____ (your company name) to make a proposal to you on the _____ project. I know that you only entertained a few proposals, and we are honored that you chose us as one of them.

Regardless of whether or not we end up doing this job, we appreciate the opportunity of bidding on it, and hope that you will also consider us for your next projects.

When time allows, I would appreciate a few moments of your time to get your assessment of our proposal. Specifically, I would like to know what you thought were its strengths and weaknesses.

Thank you again for your interest in _____ (your company name).

Sincerely,

_____

(Name)
(Title)

Lost Customer Letter

ABC CONTRACTORS

123 Any Street
Anytown, US 00000
555-555-5555

(Date)

(Name)
(Address)
(City, State, Zip)

Dear _____,

It recently came to my attention that you are no longer working with us in a business relationship. Because our business associates are as important to us as our customers, and because we have great confidence and respect for you, I ask that you take a moment to respond to this letter. I have enclosed a stamped, self-addressed envelope for your convenience. Please write and inform me of anything I did, or failed to do, that might possibly have offended you. You are highly valued by us as a business associate, and we do not wish to end our working relationship with you.

Sincerely,

_____

(Name)
(Title)

Enclosure

Bad Check Letter

ABC CONTRACTORS

123 Any Street
Anytown, US 00000
555-555-5555

(Date)

(Name)
(Address)
(City, State, Zip)

Dear _____,

This letter is to notify you that your check _____ (number), dated _____, 2007, payable to our company in the amount of $_____ has been returned by the bank as insufficient.

Although we understand that mistakes are sometimes made, we ask that you understand that we are dependent on payment from our customers in a timely manner so that we can continue to do business. Please take care of this matter immediately. Please send or bring to our offices a certified check covering the amount of the original, plus the bank charges for the overdraft, which we were required to pay. The total amount due is $_____. Your promptness will allow us to bring your account current and return your original check.

Sincerely,

_____

(Name)

(Title)

Enclosure

Collection Letter #1

ABC CONTRACTORS

123 Any Street
Anytown, US 00000
555-555-5555

(Date)

(Name)
(Address)
(City, State, Zip)

Dear _____,

After repeated attempts to collect the balance you owe on your account with us, we must inform you that we cannot extend this account any longer. You have ____ days to pay the amount that is long overdue, or our legal staff will proceed to collect this debt. If the account goes to collection, additional fees will be added to your bill.

We feel we have made every effort to work with you through repeated phone calls and written notices, all to no avail. We cannot ignore this matter any longer. We expect payment by _____ (date).

Sincerely,

_____

(Name)

(Title)

Collection Letter #2

ABC CONTRACTORS

123 Any Street
Anytown, US 00000
555-555-5555

(Date)

(Name)
(Address)
(City, State, Zip)

Dear _____,

Our recent attempts to contact you regarding payment of your overdue account have been ignored. We have offered several suggestions as to how you might bring your account current and protect your credit rating, but you must make some response soon!

Please understand, we will not simply write off this debt. We must have payment; if you call, we might possibly work out some payment arrangements, but you must call or come by our offices immediately.

Sincerely,

_____

(Name)

(Title)

Collection Letter #3

ABC CONTRACTORS

123 Any Street
Anytown, US 00000
555-555-5555

(Date)

(Name)
(Address)
(City, State, Zip)

Dear _____,

As a final attempt to collect this bill without resorting to legal action, we ask for some response to our repeated attempts to reconcile this matter. We have offered to extend the payment period, make your payments smaller, direct you to a lender, we have literally pleaded with you to discuss the matter with us. You simply do not respond.

I am sure you understand how frustrated we are; we have exhausted every resource. Our attorney recently advised us that there are several avenues available for collecting this debt, but we hesitate, even now, to go to court. Therefore, we are extending your credit for _____ days; if you do not respond by that day we have no choice but to seek legal action. We must have a check in the amount of $_____ on or before _____ (date).

Sincerely,

_____

(Name)

(Title)

Guarantee Letter

ABC CONTRACTORS

123 Any Street
Anytown, US 00000
555-555-5555

(Date)

(Name)
(Address)
(City, State, Zip)

Re: _____ (type of work) guarantee

Dear _____,

Please note that our work at the following location has been completed, and that we will guarantee this work for a period of _____ (length of time):

Owner Name: _____

Address: _____

Phone: _____

Please note that our guarantee is in accordance with standard _____ (trade designation) practices and policies.

Sincerely,

_____

(Name)

(Title)

Equipment Warranty Letter

ABC CONTRACTORS

123 Any Street
Anytown, US 00000
555-555-5555

(Date)

(Name)
(Address)
(City, State, Zip)

Re: _____ (project name or designation)

Dear _____,

We have been asked by _____ (name) of _____ (company) to put the _____ system into operation for this project. Apparently you discussed this with him also.

Just to connect the system would not pose a difficulty for us, but we do have a few concerns that need to be addressed before we can do so.

As you know, by our contract, we must guarantee all of the equipment for one year after the owner's acceptance. If we put these systems into service now, they will be in service three or four months prior to the owner's acceptance. This puts us at an extra risk. While we don't expect any problems, the warranties we get from manufacturers are for one year from the date they begin service. If something were to break down one year from today, we would be at risk. This rarely happens, but who can tell if a piece of equipment was manufactured incorrectly?

Do you think you could get us a letter signed by the owner, stating acceptance of these systems when they are put into service? If this would require an extra inspection, we will have our people prepare for it and make it as painless as possible.

I appreciate your help in this matter. We are sorry to cause a difficulty, but we just don't want to take risks.

Sincerely,

_____

(Name)
(Title)
cc:

Late Pay Letter #1

ABC CONTRACTORS

123 Any Street
Anytown, US 00000
555-555-5555

(Date)

(Name)
(Address)
(City, State, Zip)

Dear _____,

We have attempted to contact you several times by letter and/or phone to discuss payment arrangements for the balance owed on your account. The following items remain unpaid:

No.: _____ Date: _____ Amount: _____

No.: _____ Date: _____ Amount: _____

No.: _____ Date: _____ Amount: _____

At this point, we have heard nothing from you. Therefore, we are insisting that you make immediate payment. Send the amount due in the enclosed envelope.

Sincerely,

_____

(Name)

(Title)

Enclosure

Late Pay Letter #2

ABC CONTRACTORS

123 Any Street
Anytown, US 00000
555-555-5555

(Date)

(Name)
(Address)
(City, State, Zip)

Dear _____ ,

Once again we are making an attempt to collect the balance owed on your account, which is now_____ days overdue. Despite the fact that you have ignored every attempt by us to collect this debt, we feel we must remind you that we cannot keep your account open any longer without some payment.

Please call our offices today so that we can work together to reconcile this matter.

Sincerely,

_____

(Name)

(Title)

Enclosure

Late Pay Letter #3

ABC CONTRACTORS

123 Any Street
Anytown, US 00000
555-555-5555

(Date)

(Name)
(Address)
(City, State, Zip)

Dear _____,

We have not received a reply from you regarding your overdue account. However, we are sure that you will immediately remit a check in the amount of $_____ to bring your account current.

Please use the postage paid envelope that is enclosed for your convenience.

Sincerely,

_____

(Name)

(Title)

Enclosure

Late Pay Letter #4

ABC CONTRACTORS

123 Any Street
Anytown, US 00000
555-555-5555

(Date)

(Name)
(Address)
(City, State, Zip)

Dear _____,

We are confident that you will not object to a reminder that a balance of $_____ remains unpaid on your account. If you have not mailed your payment, please do so immediately, so that your account can be restored to a current status.

Cordially,

_____

(Name)

(Title)

Enclosure

Payment Received Late Letter

ABC CONTRACTORS

123 Any Street
Anytown, US 00000
555-555-5555

(Date)

(Name)
(Address)
(City, State, Zip)

Dear _____,

Thank you for sending your payment check on your recent bill.

Our collection letters proceed automatically and occasionally a payment crosses in the mail. This is obviously what happened in your case.

Your check, of course, has been properly documented and your account is currently marked paid in full.

Sincerely,

_____

(Name)

(Title)

Qualification Form Letter

ABC CONTRACTORS

123 Any Street
Anytown, US 00000
555-555-5555

(Date)

(Name)
(Address)
(City, State, Zip)

Subject: Request for Contractor Qualification Information

Dear _____,

In response to inquiries or as part of our continuing effort to identify new sources for participation in the work we perform, we have enclosed a Contractor Qualification Form which we request you to fill out and return to us.

The information regarding your firm will be kept in the strictest confidence and will be used only as a basis for evaluating your firm and your compatibility with the projects in which we will be engaged.

Kindly indicate on the form the types of contracting in which you are engaged and the geographic areas in which you work. The listing of work references and accurate indication of bonding capacity are very important.

Your completed Contractor Qualification Form will be kept on file, and will be considered when we are bidding new projects that require your specialty. Some of our clients and their architects may participate jointly with us in the selection of bid sources, and judgments do vary as to selection. Your interest in our projects is appreciated, and we look forward to receiving the completed Contractor Qualification Form from you.

Sincerely,

_____

(Name)

(Title)

Enclosure

Rate Increase Letter

ABC CONTRACTORS

123 Any Street
Anytown, US 00000
555-555-5555

(Date)

(Name)
(Address)
(City, State, Zip)

Dear _____,

There has been a rapid rise in labor and operating costs. Because of these facts, _____ (name of company) has reluctantly decided to increase service charges. This will be effective on _____, 20_____.

We appreciate your patronage and look forward to continuing our business relationship.

If you have any questions, please call us at _____.

Sincerely,

_____

(Name)

(Title)

Referral Thank You Letter

ABC CONTRACTORS

123 Any Street
Anytown, US 00000
555-555-5555

(Date)

(Name)
(Address)
(City, State, Zip)

Re: _____ (project name or designation)

Dear _____,

We were recently contacted to present a proposal to _____ (name of customer) for _____ (description of work).

We were advised that our company was referred by you, and we thank you very much for such a nice referral.

Please be assured that we will provide _____ (name of customer) with the highest quality of service and workmanship.

Thank you again for your gracious referral.

Sincerely,

_____

(Name)
(Title)

Refund Letter

ABC CONTRACTORS

123 Any Street
Anytown, US 00000
555-555-5555

(Date)

(Name)
(Address)
(City, State, Zip)

Dear _____,

It turns out that you are right.

Although I thought that our charges were justified, your letter as well as a discussion with the service crew that visited you has convinced me that you are partially correct. The _____ (item in question) must have been defective and that is probably why it has given you so much difficulty.

Therefore, I have told the bookkeeping department to cancel all of last _____ (time period's) service charges.

Now I understand that all the equipment is working smoothly. Please let me assure you that we will continue to stand by our product for a full _____ days.

Sincerely,

_____

(Name)

(Title)

CHAPTER THREE

Forms and Letters for Projects

This chapter contains useful forms, contracts, and letters written to be utilized for projects and jobs.

FORMS

Pre-Job Planning Topics

This is a job planning checklist designed primarily for a subcontractor. The last several questions are specific to an electrical contractor; you may change them as required for your company.

Project Startup Checklist

This is a project start-up checklist designed for a general contractor.

Project Management Checklist

This is a project checklist designed for a general contractor or other project manager.

Bid Receipt Log

This is a very handy form to have on bid days when phone bids are coming in fast. It contains places for the necessary information and will help you remain organized while under stress.

Bid Submission Log

This one is for noting all of your outgoing bids. On a busy day, this form can save you from a lot of pain.

Bid Form

This is a basic form used for clarifying which price pertains to which bid item. These can be very useful on a hectic bid day. Mistakes at these times can be very costly.

Bid Notes

This type of form is generally used for a formal submission of the bid.

Estimate Form

This is a very basic form for itemized estimates. Show items, quantities, material costs, and labor hours to install each item. Then multiply material and labor by quantity and add the columns.

Unit Price Estimate Form

Use this form to develop prices on a "per unit" basis.

Project Estimate Form

This is a large, detailed form for a large, complex general construction project.

FHA Cost Breakdown Form

Use this to provide cost information to the Federal Housing Administration in a form they accept.

Material Schedule

Basic form to itemize quantities and uses of materials.

Material Report

This is a form that you give to project supervisors, foremen, etc. It can be useful for job management and documentation of job issues. It is similar to a daily log, but focused on materials.

Procurement Schedule

Use this to keep track of materials on a large job where many items must be approved, ordered long in advance, and so on.

Jobsite Safety Report

This is for use during a job site inspection.

Progress Chart

Use this form to keep track of actual progress on a job. It is very useful for comparison with estimated progress. Where the two diverge, you need to find out why. This will enable you to find and correct inaccuracies between your estimates and progress on the job site.

Change Proposal Summary

This is a notice of past-due invoices and a request for immediate payment.

Change Order

This is a basic change order form. Failure to get this form properly authorized (signed by a responsible party) results in about half of all change orders going unpaid.

Daily Construction Report

This is a daily report for a construction project, generally to be filled out by the foreman of a single trade.

Daily Log

This is a daily report for a construction project, generally to be filled out by a general superintendent. It contains space for data on many trades.

Daily Worksheet

This is a basic record of what is used on a job on a particular day. Such forms can be useful, but beware of over use. (Why waste worker time on accounting tools that will never actually be used?)

Punch List

This is a basic end-of-project "punch list."

Project Close-out Checklist

This is a project close-out checklist designed for a general contractor.

Certification of Satisfaction of Lien

This certificate states that a previously filed lien has been satisfied and its demands met. Thus the lien is retired by this legal certificate.

Notice of Intention to File a Mechanic's Lien

This is a notice to a property owner that you intend to file a lien against a certain piece of property. It specifies the amount of the lien and the legal description of the property.

LETTERS

Unconditional Waiver and Release Letter

Caution: This document waives rights unconditionally and states that you have been paid for giving up those rights. This document is enforceable against you if you sign it, even if you have not been paid. If you have not yet been paid, use a conditional release form.

Conditional Waiver Final Payment Letter

This is a final payment letter, releasing the owner or contractor from all obligations to the party being paid.

Conflict Letter

Since the architectural and engineering firms are the ones that specify how the project will be constructed, they are the final arbiters on conflicts between different trades. When these situations do arise, it is your obligation to immediately resolve the situation; avoiding it will only allow the problem to grow. To solve the problem you must involve the architect or engineer or both in some cases. In such communications there are two basic rules to be followed:

1. Do not lay blame on the architect or engineer.

2. Provide a solution to the problem.

You must always ask for the specifier's approval before going ahead with any changes, but designing the changes beforehand provides advantages. First, it often makes the specifier's life a bit easier. Second, it allows you to specify your own answer to the problem. Obviously, you will have to be fair, but providing an answer often allows you to do things your own way.

Bid Letter

This is a cover letter for a bid submitted by a subcontractor to a general contractor. This is a very basic letter, but it provides a nice, professional touch when submitting a bid.

Bidder – No Plans Letter

This is a request for bid documents and is to be included in the bid. This letter can be used by a variety of companies. It can be sent from a subcontractor to a general contractor, from a general contractor to an architect, or from a general contractor to an owner.

Addendum Letter

This is a simple cover letter for a new addendum on a project that is bidding. Notice that it requests a phone call if there are any questions or difficulties. When sending addenda, make sure that you are very careful in numbering, statements of bid date and time changes, and assuring that every bidder gets a copy. This is true not only for the architect that must send out addenda, but also for the general contractor and subcontractor who must keep addenda flowing downstream without delay.

Design/Build Authorization Letter

This is the initial agreement in the design/build process. It authorizes the contractor to spend the necessary time and effort to establish a basic design and budget. The contractor will be paid for these services (the exact amount will be specified in this letter), with that amount being credited to the total design fee, should the project proceed to the construction stage. Notice that this letter is to be counter-signed by the recipient, and a copy should be returned to you.

Design/Build Cover Letter

This is a cover letter for a design/build contract proposal. It reports to the customer that you have completed the basic design for the project, which the previous letter authorized you to do. The letter then informs the customer that the next step is to prepare complete construction drawings. It also mentions that you have also completed a detailed schedule for the completion of the project by the designated date.

Design/Build Introduction Letter

This letter is written regarding a project that is being bid by the formal process. (Designed by an architect, bid to the owner according to plans and specs.) If, rather than simply bidding according to plans and specs, you find changes that could be made in the project to save money, you may gain a competitive advantage by designing the project to be less expensive, yet with the same quality standards. In this letter, you should outline the changes you wish to make, explain why they don't really affect the quality of the installation, and give the owner a figure as to how much money the changes will save.

Design/Build Modification Letter

This letter is a cover letter to the formal "plans and specs" bid. In it you identify the bid documents and acknowledge that you are bidding according to them. Then you detail the money saving design advantages that you have found and itemize their costs.

Design Estimate Letter

This is a blank proposal form for a design estimate. Note several things about it: First of all, it references the other communications you have had regarding the project. Second, it leaves you a space for filling in the design details. Explain the key parts of the design and the benefits of or reasons for the choice of these methods. Third, the form leaves you space for filling in each item on a pricing list. For a design estimate you must price each line item so that the owner can see exactly what he or she is paying for.

In many cases, an architectural firm is called upon by the owner to prepare a basic design; and then the contractor (you) is called in to design the actual implementation of the architect's generic design. In such cases, you should reference the architect and the date of the plans and specifications you worked from. This letter, as it appears here, is written in this way. If, instead, your firm has handled all of the design details from the beginning, you will need to modify this paragraph. Instead of referencing the architectural company, note that the plans, specifications, and design criteria were developed by your firm and note the date of the final design documents. After the owner reviews and modifies the estimate, you will prepare a formal contract for construction and then go to work.

Submittal Letter

This is cover letter for a set of submittals. Since one of the main concerns with submittals is that you get the approved copies returned to you promptly, this letter includes a paragraph that tells the recipient when you expect to get them back. As with many of these letters, this one quietly lays a foundation of legal evidence. Specifically, it gives you a legal framework for claiming a delay, should you not get the submittals back in time.

Hazmat Submittal Letter

The documentation and handling of hazardous materials (often shortened to "hazmat") are largely a paperwork problem. The amount of paperwork, however, depends on the types of materials that will be used in the project. Although these regulations seem rather needless, do not underestimate the power of the Environmental Protection Agency (EPA) to enforce them. Like OSHA, and even more, they can penalize your company heavily for even a minor infraction. This is a cover letter to be included with hazmat information sheets that you are supplying to your customer. Note that the letter is used to shift some of the responsibility to the recipient of the letter, or at least off of you and your company. The letter says that you believe the sheets to be in compliance with the law, but that the other party should notify you if this is not so.

General Delay Notice Letter

Because of the big threats of litigation and liquidated damages, delays (or rather, being blamed for the delays) can be a very serious liability. Letters such as the ones you find here are crucial in the avoidance of blame. Documentation almost always holds up in court—remembered conversations very rarely do. This letter is in response to delays that are being caused by others, and should usually be used by subcontractors. (They could be used by general contractors if modified.) It is especially applicable to those trades that tend to finish last (electrical, painters, carpet layers, etc.) Note that you should send this letter at the first sign of trouble brewing. This letter is essentially a preventative measure.

By sending it you are informing the recipient that:

1. He has been notified that you are not the cause of this problem and will not be held responsible.

2. You are building a legal record of your innocence in this matter, and he will not be able to win any lawsuit against you.

You are initiating the discussion, and the recipient of the letter will be reacting to your letter; which means that you are preparing the battlefield before any arrows fly.

Delay Notice Letter – Other Trade

This is very similar to the previous letter, except that it does not specify the specific cause of the delay, but instead explains the effects of the delays. This letter is not about delays your company may have caused, but about delays caused by others. There may be situations where you want to talk about the delay not being your fault, but you don't yet want to point the finger at someone else. In such cases, a letter like this will work very well. With some modification this letter could be used by a general contractor or construction manager as well as by a subcontractor.

Weather Delay Letter

This is a letter regarding delays caused by weather. This letter is a polite but firm request for a job delay allowance due to adverse weather conditions. Note that the power of this letter is that you have a foreman's log to back you up. If you don't, your evidence is thin. You can get old weather reports showing that there was rain, snow, or some other type of difficulty, but you cannot verify the actual conditions of the job site that were caused by the weather.

Job Problem Letter

This is an on-the-job form designed to document the situation and to get a speedy resolution.

Subcontractor Problem Letter

This letter is either from a general contractor to a subcontractor or from a subcontractor to a specialty contractor and is written in response to poor performance by the recipient. As with most of these letters, this one has a dual purpose—to remedy the situation and to legally protect the writer if the situation is not remedied. In such a situation, you want to be sympathetic to the other contractor's difficulties, but you can not allow them to put you, your company, and your employees at risk. At the end of the letter, further action is threatened should the situation not be immediately resolved.

Pull Off Letter #1

This is a notice of intent to pull off a job, from a subcontractor to a general contractor, from a general contractor to an owner, or from a specialty contractor to a subcontractor. It says that you are being forced to pull off of the project. Note that it references other letters that have been sent previously and says that you will have to charge for mobilization and demobilization expenses if you are forced to pull off. Note also that the letter says that you specified mobilization and demobilization on your original schedule of values. (You should always do this. The little bit of extra effort it takes will allow you a lot of negotiating room in many different circumstances.)

The final paragraph of the letter is conciliatory, without giving away any ground. You are simply sorry that the situation worked out poorly, and you would like to resolve it well. You are not, however, accepting any blame. If the situation looks bad, you should send this letter by certified, return-receipt mail.

Pull Off Letter #2

This is a very serious letter regarding a situation that could easily end up in litigation. Such a letter should be sent by certified mail so that you have proof of delivery. In this letter you will establish your legal framework by referencing the previous letters you have sent, by requesting a response, and by informing the other party that you will be seeking legal counsel. Note that the letter says that you will not be forced into accepting any risks beyond those that are specified in your contract. This is where many construction companies get themselves in trouble. Go no further than your contract mandates. You have no legal or moral obligation to go any farther than that point, and, in fact, you are jeopardizing your company, your employees, and their families if you do.

Notice of Completion Letter

This is a cover letter to be sent with the project close-out documents. In essence, it says, "Okay, we finished the project, here is the final paperwork, now pay me."

Change Order Letter

This is another type of verification of extra work. Many contractors have used this type of verification instead of a change proposal. While it is certainly better to use this rather than nothing, it would be far better to get formal approval in advance. This letter is not only a friendly confirmation of a work order, but it is also a legal statement of what you are about to do. After receiving this letter, the other company can never claim ignorance of the situation. Since most unpaid extras are justified by some type of ignorance, this letter can be very helpful. Note also that the letter says you will be billing this extra work in the same way as you do for small jobs. If this will provide a difficulty for your accounting system, change the wording of this section.

Pre-Job Planning Topics

- When are we scheduled to start?
- When are we supposed to finish?
- How is the GC to work with?
- Who are the other major subcontractors? How are they to work with? Are there any services that we can trade with them? (You let us use your backhoe, we get power into your trailer for free, etc.)
- Who is the architect and/or engineer? How are they to work with? What are their peculiarities?
- Do we expect many change orders?
- What are the temporary power requirements?
- Will we need to use any specialty subcontractors?
- When do we need to have submittals ready?
- Will we be allowed to make substitutions?
- Who are we going to buy light fixtures from? What about switchgear? How about pipe and wire?
- Can we get paid for material stored on the site?
- How much will insurance cost for this material and trailer?
- Will we rent a storage trailer, or do we have our own?
- What does the job schedule look like? Are there any obvious problem areas in the schedule?
- Will we have any problem getting deliveries at the required times?
- Will we have any problems having enough workers at any particular times?
- Will we have any problems with cash flow?
- How will payments be processed?
- What is the percentage of retainage? Will this percentage change when the job is 50% complete?
- How will change orders be handled?
- Are we locked in to any certain percentages of overhead and profit? If so, what are they?
- Does this GC usually pay on time? If not, what can we do to make him pay us on time?
- Do we need to write any letters to inform the GC or architect of problem areas?
- Who will be the inspector on this job? What will we have to do to keep him happy?
- Will we need any special equipment? If so, will we need to buy it, or can we rent it?
- What materials will we need on the job first?
- Is there any storage available, or will that be available later in the building?
- Which of the job foremen would be best suited for this particular project? Will he be available, or do we need to choose someone else?
- Are there any particular electricians that would do especially well on this job?
- How good are the blueprints?
- Should we make any of our own drawings to help the electricians?
- Will they need details drawn for any specific areas or rooms?
- Can we redesign the electrical system at all?
- Can we combine two or three circuits in one raceway?
- Are there any inadequacies in the plans or specs?
- Who supplies the motor starters?
- Who does the control wiring?
- Who provides disconnect switches for mechanical units?
- Who is responsible for pitch pans and sealing roof penetrations?
- Are there any particular items or assemblies that we could pre-assemble in our shop?
- Are there any new techniques or tools that would work well on this job?

Project Startup Checklist

_____   Define project organization.

_____   Review contract.

_____   Define scope of work.

_____   Execute all subcontracts.

_____   Meet with all subcontractors.

_____   Meet with the owner.

_____   Meet with architect and engineers.

_____   Kickoff meeting.

_____   Define purchasing procedures.

_____   Set budget.

_____   Verify insurance requirements.

_____   Write schedule.

_____   Cash flow projection.

_____   Prepare project files.

_____   Prepare project distribution list.

_____   Obtain building permit.

_____   Prepare site crew.

_____   Site kickoff meeting.

_____   Review drawings and specifications.

_____   Verify availability of tools and machinery.

_____   Verify and document site conditions.

_____   Verify physical boundaries.

_____   Plan site logistics.

_____   Review tests, inspections, and quality control procedures.

_____   Post safety programs.

_____   Post emergency phone numbers.

_____   Prepare temporary facilities.

_____   Prepare site security.

_____   Install project sign.

_____   Prepare all submittals.

_____   _____

_____   _____

_____   _____

_____   _____

_____   _____

_____   _____

_____   _____

_____   _____

_____   _____

_____   _____

_____   _____

File Type: Word | **File Name:** 62-Project-Startup-List.doc

Project Management Checklist

ABC CONTRACTORS

123 Any Street
Anytown, US 00000
555-555-5555

Date: _____   Contractor: _____

Owner: _____   Project: _____

	Estimator	Project Mgr.	Supervisor	Foreman
1. Submit job info.				
2. Determine the basic organization of the project.				
3. Assign key employees.				
4. Review the contract.				
5. Define scope of work.				
6. Define procurement authority.				
7. Establish the project accounting code system.				
8. Get approval for the budget.				
9. Make sure all departments are aware of the project and are prepared to handle the work load.				
10. Get safety insurance requirements.				
11. Prepare project and procurement schedules.				
12. Determine anticipated cash flow.				
13. Set up office and job-site files.				
14. Publish the project meeting schedule.				
15. Hold a pre-construction conference with appropriate parties.				

File Type: Excel | **File Name:** 63-Project-Management.xls

Project Management Checklist (*cont.*)

	Estimator	Project Mgr.	Supervisor	Foreman
16. Establish communication channels among project participants.				
17. Resolve drawing and specification questions.				
18. Designate the field crew.				
19. Reserve appropriate materials, tools, and equipment from the company inventory, if any.				
20. Conduct a visual and photographic survey of existing conditions. Verify condition of adjacent off-site items.				
21. Verify physical boundaries.				
22. Determine site logistics.				
23. Prepare job site accident prevention program.				
24. Post the job site accident prevention program, safety posters, and emergency contact numbers.				
25. Arrange for temporary facilities.				
26. Review quality control procedures, including required tests and inspections.				
27. Review the job site security.				
28. Install the project sign.				
29. Obtain a building permit.				

File Type: Excel | **File Name:** 63-Project-Management.xls

Bid Receipt Log

Date: _____

Project
Designation: _____

Address: _____

_____

_____

Bid documents:

_____

_____

_____

Bidder	Bid Amount	Date/Time	Comments
_____	_____	_____	_____
_____	_____	_____	_____
_____	_____	_____	_____
_____	_____	_____	_____
_____	_____	_____	_____
_____	_____	_____	_____
_____	_____	_____	_____
_____	_____	_____	_____
_____	_____	_____	_____
_____	_____	_____	_____
_____	_____	_____	_____
_____	_____	_____	_____
_____	_____	_____	_____
_____	_____	_____	_____
_____	_____	_____	_____
_____	_____	_____	_____

Bid Submission Log

Date: _____

Project
Designation: _____

Address: _____

_____

_____

Bid documents:

_____

_____

_____

Bid To	Bid Amount	Date/Time	Comments

Bid Form

ABC CONTRACTORS

123 Any Street
Anytown, US 00000
555-555-5555

Contact: _____   Date: _____

Phone: _____   Project: _____

Email: _____

Item #: _____

Bid Price: _____

Includes: _____

_____

_____

Excludes: _____

_____

Alternate: _____

Alternate Price: _____   [ ] included in price above   [ ] not included

Item #: _____

Bid Price: _____

Includes: _____

_____

_____

Excludes: _____

_____

Alternate: _____

Alternate Price: _____   [ ] included in price above   [ ] not included

Item #: _____

Bid Price: _____

Includes: _____

_____

_____

Excludes: _____

_____

Alternate: _____

Alternate Price: _____   [ ] included in price above   [ ] not included

Bid Notes

Bid No.:		Date:

Job:	
Location:	
Company:	
Address:	Prepared By:
City/State/Zip:	E-mail:
Type of Work:	Phone:

WORK	**BID**
BID TOTAL	

EXCLUSIONS AND QUALIFICATIONS

ACKNOWLEDGEMENT OF ADDENDUM	**Tax**	
DELIVERY	Excluded	
	Included	
Received By:		

File Type: Excel | **File Name:** 67-Bid-Notes.xls

Date:_____

Estimate

Building:	Prepared By:
Location:	Approved By:
Architects:	Estimate No.:

Item	Quantity	Material	Extension	Labor Hours	Cost
			Subtotals		

Notes:
_____
_____
_____

Unit Price Estimate Sheet

ABC CONTRACTORS

123 Any Street
Anytown, US 00000
555-555-5555

Date: _____

Estimate No.: _____

Project: _____

Task Description	Materials				Labor				Sub Bids/ Equip. Rental	Task Total	Notes/ Allowances
	Unit	Amt.	Cost	Total	Unit	Amt.	Cost	Total			
Subtotals:											

Contingency _____ % _____

Overhead & Profit _____ % _____

Total Estimate _____

File Type: Excel | **File Name:** 69-Unit-Price-Estimate.xls

Project Estimate Form

Description	Materials			Labor				Sub Bids	Total
	Unit	Amt.	Cost	Total	Unit	Amt.	Cost	Total	Equipment	
FEES										
City License										
Coastal Development										
Dedication										
Environmental Health										
Front End Advances (Utilities)										
Miscellaneous										
Mitigation										
Parking										
Parks and Recreation										
Plan Check (General)										
Safety/OSHA										
School District										
Sewer Assessment										
Traffic Control										
Transportation Corridor										
Utilities Tie-in										
Water Department—Meters										
Water Share Rights										
PERMITS										
Building										
Demolition										
Electric										
Encroachment										
Fireplace										
Fire Sprinklers										
Grading										
Mechanical										
Miscellaneous										
Plumbing										
Power Pole—Temporary										
Public Work										
Retaining or Property-line Walls										
Sewer Cap										
Street Use/Dumpster/etc.										

File Type: Excel | **File Name:** 70-Project-Estimate.xls

Project Estimate Form (cont.)

Description	Materials				Labor				Sub Bids	Total
	Unit	Amt.	Cost	Total	Unit	Amt.	Cost	Total	Equipment	
TEMPORARY										
Barricades										
Carpenter										
Container Storage										
Crane										
Day Labor										
Dewatering										
Dumpster										
Dust Protection										
Electric Meter Billing										
Fencing										
Finish Carpenter										
Job Shed or Trailer										
Job Toilet										
Miscellaneous										
Phone/Fax/Computer/DSL										
Power Pole and Equipment										
Progress Photos										
Protection—Finishes										
Protection—Pedestrian Walk with Cover										
Protection—Weather										
Security Lighting										
Semi-Skilled Labor										
Superintendent—Full/Part-time										
Transportation										
Water Billing										
INSURANCE										
Builder's Risk—Fire and Theft										
Performance Bond										
Public Liability and Completed Operations										
MISCELLANEOUS										
Architectural Drawings										
As-built Drawings										
Bid Procedure										

File Type: Excel | **File Name:** 70-Project-Estimate.xls

Project Estimate Form (cont.)

Description	Materials				Labor				Sub Bids	
	Unit	Amt.	Cost	Total	Unit	Amt.	Cost	Total	Equipment	Total
Cleanup Final Subcontract										
Engineering Consultants										
General Conditions—Miscellaneous										
Inspectors										
Operation and Maintenance Manuals										
Prints and Specifications										
Product Warranties										
Supplies—Billable to Job										
DEMOLITION										
Asbestos Removal										
Concrete Coring										
Cut, Break, and Remove All										
Indicated Concrete										
Major Demolition										
Shoring										
Tree Removal										
GRADING										
Backfill and Compact										
Backhoe Trench										
Erosion Control and Drainage										
Footing Soils Removal										
Grading—Finish										
Grading—Rough										
Import/Export										
BUILDING UTILITIES										
Cable TV Service										
Electrical Service										
Fire Hydrant										
Fire Sprinkler Service										
Gas Service										
Laterals Sewer										
Relocate/Remove Power Pole										
Relocate/Remove Street Light										
Storm Drainage										
Street Signs										

File Type: Excel | **File Name:** 70-Project-Estimate.xls

Project Estimate Form (cont.)

Description	Materials				Labor				Sub Bids	
	Unit	Amt.	Cost	Total	Unit	Amt.	Cost	Total	Equipment	Total
Telephone Service										
Water Service										
ASPHALT										
Patch										
Paving										
Slurry Coat										
Stripping and Bumper										
GATES AND FENCES										
Fence—Chain Link/Wood/Glass										
Gates—Chain Link/Wood/Glass										
Wrought Iron or Metal										
LANDSCAPING										
Backflow Preventer										
Landscape Lighting										
Planting and Irrigation										
Planters—Preformed										
Stepping Stones										
CONCRETE										
Footings										
Gunite										
Offsite—Sidewalk/Curb/Gutter/Driveway										
Onsite—Swale/Curb/Sidewalk/etc.										
Piles and Caissons										
Precast Products										
Retaining Wall—Poured in Place										
MASONRY										
Fencing or planters										
Retaining walls										
Trash enclosure										
MISCELLANEOUS SITEWORK										
Bollards										

File Type: Excel | **File Name:** 70-Project-Estimate.xls

Project Estimate Form (cont.)

Description	Materials				Labor				Sub Bids	
	Unit	Amt.	Cost	Total	Unit	Amt.	Cost	Total	Equipment	Total
Bridge										
Cap Sewer										
Civil Engineering and Survey										
Docks and Boat Facilities										
Drainage Systems										
Painting—Gates/Fence/Bollards/Other										
Shoring and Bulkheading										
Sitework—Miscellaneous										
Soil Poisoning or Treatment										
Soils Engineering and Inspection										
Stucco—Walls/Trash Area/Planter/etc.										
Vaults and Vault Doors										
CONCRETE										
Concrete Patch Work										
Foundation—Footings and Slab										
Gypcrete, Lightweight, etc.										
Lab Testing Concrete										
Prestress Concrete Construction										
Tilt-up Construction										
MASONRY										
Block Foundation Walls										
Glass Block										
Lab Testing Masonry										
Masonry Special Finishes—										
Cast Concrete Color										
Miscellaneous										
Sandblasting										
Stonework										
Veneers										
METAL										
Equipment Screens										
Light Gauge Frame										
Ornamental—Handrails/Railings/etc.										
Shop Drawings										
Siding										
Stairs/Spiral/Access Ladders										
Steel—Miscellaneous										
Structural Connections										

File Type: Excel | **File Name:** 70-Project-Estimate.xls

Project Estimate Form (cont.)

Description	Materials				Labor				Sub Bids	
	Unit	Amt.	Cost	Total	Unit	Amt.	Cost	Total	Equipment	Total
Structural Steel										
Structural Steel Specialties										
CARPENTRY										
Architectural Millwork										
Cabinets—Millwork Installed										
Carpenter Rough Set Exterior										
Doors and Windows										
Carpentry Finish										
Carpentry Rough and Light Hardware										
Fireplace Mantel										
Light Framing Hardware										
Lumber—Finish										
Lumber—Glue Laminated Beams										
Lumber—List (Send for Builder's List)										
Lumber—Rough										
Lumber—Truss or Joist										
Lumber—Wood Siding										
Plastic—Laminates										
Plastic—Special Fabrications										
Special Railings										
Storage Shelving										
Wood Base and Casing—Material										
Wood Siding Labor to Install										
MOISTURE PROTECTION										
Caulking and Sealants										
Insulation/Air Infiltration										
Hot Mop—Tub and Shower Base										
Rain Gutters										
Roof—Accessories										
Roof—Patch										
Roofing										
Roofing—Metal Fabrication										
Sheet Metal—Special Fabrication										
Sheet Metal—Standard										
Skylights										
Waterproofing										
Weatherstripping										

Project Estimate Form (cont.)

Description	Materials				Labor				Sub Bids	
	Unit	Amt.	Cost	Total	Unit	Amt.	Cost	Total	Equipment	Total
DOOR, WINDOW, GLASS										
DOORS										
Access Doors										
Coil Roll up Door										
Entry Doors and Frames										
Fire Doors and Frames										
French Doors and Frames										
Garage Door										
Metal Doors and Frames										
Powered Operators/Garage Door Openers										
Security Iron Type—Rolling or Overhead										
Specialty Doors and Frames										
Wood Doors and Frames										
GLASS										
Bath Enclosure										
Mirror										
Screen Walls										
Special Application										
Stained/Beveled/Etched										
Storefront										
Wardrobe—Mirror/Vinyl/Other										
Window Glazing										
Windows—Aluminum										
Windows—Wood										
Windows and French Doors — Install										
HARDWARE										
Finish—Knobs/Latches/Closers										
Installation of Hardware—Labor										
Medicine Cabinets										
Toilet and Bath Accessories										
FINISH										
DECKING										
Elastomeric										
Hot Mop										
Wood Decking										

File Type: Excel | **File Name:** 70-Project-Estimate.xls

Project Estimate Form (cont.)

Description	Materials				Labor				Sub Bids	
	Unit	Amt.	Cost	Total	Unit	Amt.	Cost	Total	Equipment	Total
DRYWALL										
Drywall and Metal Studs—Hang, Tape and Texture										
Hang, Tape, and Texture										
Metal Stud Framing										
Special Finish										
FLOORING										
Carpet and Pads										
Floor Preparation—Subfloor/Float/Other										
Vinyl—Base										
Vinyl—Flooring										
Wood—Flooring										
PAINTING										
Exterior Only										
Interior and Exterior										
Interior Only										
Paperhanging										
Special Finishes										
PLASTER										
Exterior Only										
Interior and Exterior										
Interior Only										
Tile Backing										
TILE										
Marble										
Stone										
Terrazzo										
Tile										
SPECIAL FINISH										
Acoustical Ceilings										
Acoustical Treatments										
Corian Top										
Marlite										

Project Estimate Form (cont.)

Description	Materials				Labor				Sub Bids	
	Unit	Amt.	Cost	Total	Unit	Amt.	Cost	Total	Equipment	Total
Paneling										
Simulated Marble Top										
Unilav Top with Complete Molded Sink										
Upholstered Treatments										
SPECIALTIES										
Access Floors and Walls										
Awnings										
Attic Access Stair										
Chutes—Laundry/Trash										
Directories										
Fireplace—Masonry										
Fireplace—Prefab Metal										
Louvers and Vents										
Luminous Ceilings										
Pest Control										
Postal Facilities										
Screens										
Security Grilles										
Signage and Graphics										
Shutters—Exterior										
Stairs Construction										
Toilet Partitions										
Wardrobe and Closet Specialties										
EQUIPMENT										
Appliance										
Central Vacuum										
Miscellaneous										
Safe or Vault										
FURNISHINGS										
Blinds, Shades, and Shutters—Interior										
Drapery and Curtain Hardware										
Interior Plants and Planters										
Miscellaneous										
Moveable Partitions										
SPECIAL										
Equipment Shed or Vault										
Fountains or Waterscape										

File Type: Excel | **File Name:** 70-Project-Estimate.xls

Project Estimate Form (cont.)

Description	Materials			Labor				Sub Bids	
	Unit	Amt.	Cost	Total	Unit	Amt.	Cost	Total	Equipment	Total
Pool or Spa Decking										
Sauna and Equipment										
Swimming Pool and Equipment Shed or Vault										
Tennis Courts										
Wine Storage Room										
CONVEYING										
Dumbwaiter										
Elevators										
Wheelchair lift										
MECHANICAL										
HVAC										
Exhaust Fan System										
Heating and Air Conditioning										
Ducts—All Ventilating Fans										
Fan—Special Use										
HVAC Shop Drawings										
Registers, Grilles, and Diffusers										
Refrigeration										
FIRE EQUIPMENT										
Alarm Special Application										
Fire Sprinkler—System										
Fire Sprinkler—Shop Drawings										
Extinguishers—Cabinets										
Halon System										
Hose Cabinet—Racks, EEELs, Hose										
PLUMBING										
Cesspool/Septic Tank/Pump										
General Plumbing										
Fixtures—Fiberglass Tub/Shower										
Fixtures—Jacuzzi tub										
Fixtures—Plumbing										
Fixtures—Trim										
Shop Drawings										
Solar System										
Sump Pump/Sewer Injector										

File Type: Excel | **File Name:** 70-Project-Estimate.xls

Project Estimate Form (cont.)

Description	Materials				Labor				Sub Bids	
	Unit	Amt.	Cost	Total	Unit	Amt.	Cost	Total	Equipment	Total
Wall Heaters										
ELECTRICAL										
Alarm										
Door Bell										
Electric—General										
Exhaust Fans Interior—Bath/Kitchen/Laundry										
Shop Drawings										
Emergency Lights										
Fixtures and Lamps										
Low Communications										
Music System										
Parking Lot Lighting										
Telephone Prewire										
TV										
SUMMARY										
Carpentry										
Concrete										
Conveying										
Doors, Windows, and Glass										
Electrical										
Equipment										
Finish										
Furnishings										
General Conditions										
Masonry										
Mechanical										
Metal										
Moisture Protection										
Sitework										
Special Construction										
Specialties										

File Type: Excel　|　**File Name:** 70-Project-Estimate.xls

FHA Cost Breakdown Form

ABC CONTRACTORS

123 Any Street
Anytown, US 00000
555-555-5555

Date:_____

Sponsor:_____   Project: _____

Building ID_____   Location: _____

L	D	Trade Item	Cost	Description
1	3	Concrete		
2	4	Masonry		
3	5	Metals		
4	6	Rough Carpentry		
5	6	Finish Carpentry		
6	7	Waterproofing		
7	7	Insulation		
8	7	Roofing		
9	7	Sheet Metal		
10	8	Doors		
11	8	Windows		
12	8	Glass		
13	9	Lath and Plaster		
14	9	Drywall		
15	9	Tile Work		
16	9	Acoustical		
17	9	Wood Flooring		
18	9	Resilient Flooring		
19	9	Painting and Decorating		
20	10	Specialties		
21	11	Special Equipment		
22	11	Cabinets		
23	11	Appliances		
24	12	Blinds and Shades, Artwork		
25	12	Carpets		
26	13	Special Construction		
27	14	Elevators		
28	15	Plumbing and Hot Water		

File Type: Excel | **File Name:** 71-FHA-Cost-Breakdown.xls

FHA Cost Breakdown Form (*cont.*)

L	D	Trade Item	Cost	Description			
29	15	Heating and Ventilation					
30	15	Air Conditioning					
31	16	Electrical					
32		**Subtotal (Structures)**					
33		Accessory Structures					
34		**Total (Lines 32 & 33)**					
35	2	Earthwork					
36	2	Site Utilities					
37	2	Roads and Walks					
38	2	Site Improvements					
39	2	Lawns and Planting					
40	2	Unusual Site Conditions		**Non-Residential and Special Exterior Land Improvement**		**Offsite Costs** (Costs not included in trade item breakdown)	
41		**Total Land Improvements**					
42		**TI Structure and Land Imp.**		Description	Est. Cost	Description	Est. Cost
43	1	General Requirements					
44		**Subtotal (Lines 41 and 42)**					
45		Builder's Overhead					
46		Builder's Profit		**TOTAL: $**			
47		**Subtotal (Lines 44 thru 46)**		**Other fees:**		**TOTAL: $**	
						Demolition (Costs not included in trade item breakdown)	
48		Other Fees					
49		Bond Premium				Description	Est. Cost
50		**Total for All Improvements**					
51		Builder's Profit Paid by Means Other Than Cash					
52							
53		**Total All Improv. (less #52)**		**TOTAL: $**		**TOTAL: $**	

Mortgagor: _____ By: _____ Date: _____

Contractor: _____ By: _____ Date: _____

FHA: _____ Date: _____
Processing Analyst

FHA: _____ Date: _____
Chief Underwriter

FHA: _____ Date: _____
Chief, Cost Branch or Cost Analyst

File Type: Excel | **File Name:** 71-FHA-Cost-Breakdown.xls

Material Schedule

ABC CONTRACTORS

123 Any Street
Anytown, US 00000
555-555-5555

Type:_____   Date: _____

Project:_____   Revised: _____

Item/Description	Location									

Material Report

Date or dates: _____

Project
Designation: _____

Address: _____

_____

_____

MATERIAL: AMOUNT: WHERE INSTALLED:

_____   _____   _____

_____   _____   _____

_____   _____   _____

_____   _____   _____

_____   _____   _____

_____   _____   _____

_____   _____   _____

_____   _____   _____

_____   _____   _____

_____   _____   _____

_____   _____   _____

_____   _____   _____

_____   _____   _____

_____   _____   _____

_____   _____   _____

_____   _____   _____

DELIVERIES:

_____

_____

_____

_____

_____

_____

Procurement Schedule

ABC CONTRACTORS

123 Any Street
Anytown, US 00000
555-555-5555

Date: _____

Contractor: _____   Project: _____

Description	Submittals		Order Date	Shipping Time	Delivery	Call To Confirm	Rec'd	Sched. Instl.
	Received	Approved						

Job Site Safety Report

ABC CONTRACTORS

123 Any Street
Anytown, US 00000
555-555-5555

Date: _____   Project: _____

Contractor: _____   Inspected By: _____

Personal Protection:	**Yes**	**No**	**Notes**
1. Hard hats worn by all employees and visitors?	_____	_____	_____
2. Hearing protective devices worn by workers where noise levels are high?	_____	_____	_____
3. Eye and face protection worn by workers exposed to potential eye or face injury?	_____	_____	_____
Housekeeping:			
4. Toilet facilities maintained?	_____	_____	_____
5. Sidewalks free of construction materials and debris?	_____	_____	_____
6. Passageways and stairs clear of debris?	_____	_____	_____
7. Adequate illumination in work areas and stairways?	_____	_____	_____
General:			
8. Current employee injury logs?	_____	_____	_____
9. Fire extinguishers present?	_____	_____	_____
10. Job site accident prevention program posted?	_____	_____	_____
11. Official OSHA notice poster posted?	_____	_____	_____
12. Emergency telephone numbers posted?	_____	_____	_____
13. First-aid kit present?	_____	_____	_____
14. Warning signs posted near hazardous work areas?	_____	_____	_____

Job Site Safety Report (cont.)

Floors and Wall Openings:

15. Barrier guards provided where there is a drop no more than 4 feet at a wall opening? _____ _____ _____

16. Floor and roof edges 4 feet or more above floor or ground level provided with railings? _____ _____ _____

Scaffolding:

17. Railings provided when a platform is 7½ feet or more above floor or ground level? _____ _____ _____

18. Scaffolding set plumb with adequate foundation bearing plates? _____ _____ _____

19. Scaffold guyed or tied to structure? _____ _____ _____

20. Platform planks laid tightly together to prevent material and tools from falling through? _____ _____ _____

21. Platform planks extend over end supports not less than 6 inches? _____ _____ _____

Ladders:

22. Proper-height ladders in use? _____ _____ _____

23. Ladders in good condition? _____ _____ _____

24. Access ladders extend 3 feet beyond floor or roof secured against slipping? _____ _____ _____

Machinery and Tool Guarding:

25. Pulley belts and wheels enclosed with guards? _____ _____ _____

26. Cut-off saws provided with automatic blade returns? _____ _____ _____

27. Saw blade guards in place and operating properly? _____ _____ _____

Welding:

28. Cylinders secured to prevent being knocked over? _____ _____ _____

29. Proper eye protection worn by welders? _____ _____ _____

File Type: Excel | **File Name:** 75-Jobsite-Safety-Report.xls

Job Site Safety Report (cont.)

30. Cylinder valves closed when work
    is finished?   _____ _____ _____

31. Valve protection caps in place except
    when cylinders are in use?   _____ _____ _____

32. Cylinders in upright position while in use?   _____ _____ _____

Electrical:

33. Differing receptacles and attachment
    plugs not used?   _____ _____ _____

34. Cords out of vehicle paths?   _____ _____ _____

35. Male plug ends unmodified?   _____ _____ _____

36. Cords free of cuts?   _____ _____ _____

Other:

_____

_____

_____

_____

_____

_____

_____

_____

_____

_____

_____

_____

Comments:

_____

_____

_____

_____

By: _____   Copy to: _____

Progress Chart

ABC CONTRACTORS

123 Any Street
Anytown, US 00000
555-555-5555

Week ending: _____

Project: _____

No.: _____

Description	Est. start	Act. start	% comp	10%	20%	30%	40%	50%	60%	70%	80%	90%	100%

Change Proposal Summary

Project: _____

Customer Change Request #: _____

Contract #: _____

Prime Contractor Request #: _____

Contractor Job #: _____

Contractor Change Proposal #: _____

Description of Change: _____

1. Material . . . . $_____
2. Sales Tax _____ % . . . . $_____
3. Labor . . . . $_____
4. Labor Supervision . . . . $_____
5. Travel . . . . $_____
6. Workmen's Compensation, Insurance, and
   Payroll Taxes . . . . $_____
7. Subtotal . . . . $_____
8. Other Job Expenses . . . . $_____
9. Subcontract Cost . . . . $_____
10. Total Prime Cost . . . . $_____
11. Overhead _____% of line 10 . . . . $_____
12. Total Cost . . . . $_____
13. Profit _____% of line 12 . . . . $_____
TOTAL PRICE OF CHANGE . . . . $_____
TIME EXTENSION REQUIRED . . . . _____ DAYS
ABOVE PRICES SUBJECT TO CHANGE IF NOT ACCEPTED BY _____ (Date)

Accepted by: _____

Company: _____

Signature: _____

Date: _____

Change Order

Date: _____

Project: _____

Address: _____

_____

_____

Owner/Contractor/Subcontractor:

_____

_____

Original Contract Date: _____

Change Order No.: _____

Change the work to be performed under the original contract as follows:

_____

_____

_____

Price to be adjusted as follows:

Original contract price	$	_____
Net amount previous change orders	$	_____
Current contract price	$	_____
Adjustment this Change Order	$	_____
Revised contract price	$	_____

Schedule adjustments required by this change order:

_____

_____

Adjustments to invoicing schedule:

_____

_____

Sincerely,

_____

(Name)

(Title)

Daily Construction Report

Date: _____

Project Designation: _____

Owner: _____

Address: _____

                _____

                _____

Weather: _____

_____

_____

Work Completed: _____

_____

_____

_____

_____

_____

_____

_____

_____

_____

_____

_____

_____

_____

_____

Complications and/or problems:

_____

_____

_____

_____

_____

_____

_____

_____

_____

_____

_____

Daily Log

Date: _____

Project Designation: _____

Owner: _____

Address: _____

_____

_____

Weather: _____
_____
_____

Superintendent: _____

TRADES	WORKERS	NOTES
Bricklayers	_____	_____
Carpenters	_____	_____
Cement Masons	_____	_____
Electricians	_____	_____
Iron Workers	_____	_____
Laborers	_____	_____
Operating Engineers	_____	_____
Plumbers	_____	_____
Pipe Fitters	_____	_____
Sheet Metal	_____	_____
Roofers	_____	_____
Drywall	_____	_____
Painters	_____	_____
Flooring	_____	_____
_____	_____	_____
_____	_____	_____
_____	_____	_____
_____	_____	_____
_____	_____	_____
_____	_____	_____
_____	_____	_____
TOTAL	_____	_____

Daily Log (*cont.*)

EQUIPMENT   **HOURS**

_____   _____
_____   _____
_____   _____
_____   _____
_____   _____
_____   _____

CHANGE ORDERS   **AUTHORIZED**   **AMOUNT**

_____   _____   _____
_____   _____   _____
_____   _____   _____
_____   _____   _____
_____   _____   _____

OTHER NOTES

_____
_____
_____
_____
_____
_____
_____
_____
_____
_____
_____
_____
_____
_____
_____
_____
_____
_____

Daily Work Sheet

ABC CONTRACTORS

123 Any Street
Anytown, US 00000
555-555-5555

Owner: _____

Change Order #: _____   Project: _____

Date: _____   Location: _____

LABOR						
Remarks		Labor Classification				Weather
						Temperature
						A.M.
						P.M.

Name	Description of Work	Classification			Hours	Rate	Amount

Labor Burden _____

Total Labor _____

MATERIALS USED					
Quantity	Item	Units	Amount	Total	Place of Purchase

Total Material _____

Total Labor and Material _____

Overhead Expense _____

TOTAL _____

File Type: Excel | **File Name:** 81-Daily-Work-Sht.xls

Punch List

ABC CONTRACTORS

123 Any Street
Anytown, US 00000
555-555-5555

Owner: _____

Date: _____   Project: _____

Item No.	Description	Date Complete	Owner's Approval (Initial)
1.			
2.			
3.			
4.			
5.			
6.			
7.			
8.			
9.			
10.			
11.			
12.			
13.			
14.			
15.			
16.			
17.			
18.			
19.			
20.			
21.			
22.			
23.			

Contractor: _____   Owner: _____

Project Close-out Checklist

Date: _____

Project
Designation: _____

Address: _____

_____

_____

_____ All punch list items

_____ Final utility and service connections

_____ Acceptance checklists

_____ Final agency inspection

_____ File Notice of Completion

_____ Clean project

_____ Acceptance walk-through with the contractors and with the client

_____ Process final payment from client or lender

_____ Assist systems start-up

_____ Obtain warranties and lien releases from all subs

_____ Assemble owner's manual with warranties and project literature

_____ Assemble as-built drawings

_____ Assemble maintenance and operating instructions

_____ Initiate preventive maintenance program as appropriate

_____ Train operating and maintenance staff as needed

_____ Assemble and store spare parts and materials as required

_____ Remove excess materials

_____ Remove temporary facilities, tools, and equipment

_____ Correct actual schedule and job costs for future reference

_____ Sign-off for all project keys

_____ Obtain letter of recommendation from client

_____ Follow up with client 30 days after close of job

_____ _____

_____ _____

_____ _____

_____ _____

_____ _____

_____ _____

_____ _____

_____ _____

_____ _____

_____ _____

File Type: Word | **File Name:** 83-Project-Closeout.doc

Certification of Satisfaction of Lien

Whereas, on _____ (date), _____ (lien holder) caused a lien to be filed for record in the office

of _____ (name of office) of _____ County, State of _____, which was duly recorded in

Volume _____, page _____, of _____, against property owned by _____(owner), of

_____ (address), for _____ (description of work performed) by

_____ (name of lien holder), the description of which is as follows:

Whereas, on _____ (date), _____ (owner) fully satisfied the indebtedness secured

by said mechanic's lien.

Therefore, in consideration of said payment, _____ (lien holder) hereby certifies that

said lien is released and discharged, and directs the County _____ (name of office) to

discharge of record said lien.

_____
(Date)   _____
                                (Signature)

Notice of Intention to File a Mechanic's Lien

TO: (Name of Owner)

           (Address)

           (City, State, Zip)

Notice is hereby given that the undersigned, _____ (name) intends to file a

mechanic's lien for _____ (amount) Dollars ($), on real property owned by you and commonly

known as _____ (street address). The legal description of said property is as follows:

(legal description)

The filing of said lien, pursuant to _____ (cite statute), is for the purpose of securing payment

of amounts due for _____ (services) performed by the undersigned within the last _____

(number) days, in accordance with the _____ (written or oral) agreement entered into on

_____ (date) between you and the undersigned.

_____   _____

(Date)   (Signature)

File Type: Word | **File Name:** 85-Intention-to-File-Lien.doc

Unconditional Waiver and Release Letter

ABC CONTRACTORS

123 Any Street
Anytown, US 00000
555-555-5555

(Date)

(Name)
(Address)
(City, State, Zip)

Re: _____ (project name or designation)

Project address: _____

_____

_____

The undersigned has been paid in full for all labor, services, equipment or material furnished to on the job of _____ (owner) at _____ (job location) and does hereby waive and release any right to any mechanic's liens, stop notices, or any rights against a labor and material bond on the job, except for disputed claims for extra work in the amount of $_____ (amount) I/We certify that all labor and/or laborers have been paid in full to the below referred date.

ACCEPTED BY:

_____   _____

(Name)   (Date)

(Title)

Conditional Waiver and Release Letter

ABC CONTRACTORS

123 Any Street
Anytown, US 00000
555-555-5555

(Date)

(Name)
(Address)
(City, State, Zip)

Re: _____ (project name or designation)

Project address: _____

_____

_____

This is a receipt by the undersigned of a check from _____ (general contractor or owner) in the amount of $_____ (amount) payable to _____ (subcontractor). When the check has been properly endorsed and has been paid by the bank upon which it is drawn, this document shall become effective to release any mechanic's liens, stop notices, or bond rights the undersigned has in connection with the work being performed for _____ (owner) at _____ (job location).

This release covers the final payment to the undersigned for all labor, services, equipment, or material furnished in connection with the work, except for disputed claims for additional work in the amount of $ _____ (amount, if any.)

I/We certify that all labor and/or laborers have been paid in full by the referred date below.

ACCEPTED BY:

_____   _____

(Name)   (Date)

(Title)

Conflict Letter

ABC CONTRACTORS

123 Any Street
Anytown, US 00000
555-555-5555

(Date)

(Name)
(Address)
(City, State, Zip)

Re: _____ (name of project)

Dear _____,

I am writing regarding some concerns about a development on the job site of the project referenced above; our concerns involve the _____ (name of trade) with whom we have a conflict. Since your company is _____ (contractor, architect, or consulting engineer) we are asking you to mediate this matter. Specifically, this difficulty involves: (describe matter in detail)

_____
_____
_____

We suggest the following solutions: (describe solution)

_____
_____
_____

Please take time to review this situation personally. Our foreman _____ (name) will be glad to assist you; you may call him at _____ to make arrangements to come. Because the project is at a critical stage, we ask that you make a special effort to review the situation as soon as possible.

Thank you for your help; if you have any questions or requests, please contact us.

Sincerely,

_____

(Name)
(Title)

Bid Letter

ABC CONTRACTORS

123 Any Street
Anytown, US 00000
555-555-5555

(Date)

(Name)
(Address)
(City, State, Zip)

Re: _____ (project name or designation)

Dear _____,

Enclosed you will find our bid on the subcontract for the _____ work on the project mentioned above.

If there are any clarifications or additions you need, please let us know, and we will get them to you promptly.

Thank you for your consideration.

Sincerely,

_____

(Name)

(Title)

Bidder – No Plans Letter

ABC CONTRACTORS

123 Any Street
Anytown, US 00000
555-555-5555

(Date)

(Name)
(Address)
(City, State, Zip)

Dear _____,

Thank you for your interest in bidding the _____ project to us.

Unfortunately, we have a limited number of sets of plans, and cannot spare any at this time. If you are able to obtain plans from another source, we would be glad to entertain your bid.

Thank you again for your interest.

Sincerely,

_____

(Name)

(Title)

Addendum Letter

ABC CONTRACTORS

123 Any Street
Anytown, US 00000
555-555-5555

(Date)

(Name)
(Address)
(City, State, Zip)

Re: _____ (project name or designation)

Dear _____:

Enclosed you will find addendum #___ for the _____ project you are bidding to us.

Please note the following pertinent facts: _____
_____ (detail changes caused by the addendum)

As you review this addendum, you will find that these changes and others may affect your bid to us. Please incorporate the changes into your final bid to us which is due _____ (date).

If any of these changes are difficult to deal with, please call me immediately at _____ (phone number) or email me at _____ (email address).

Sincerely,

_____

(Name)

(Title)

Design/Build Authorization Letter

ABC CONTRACTORS

123 Any Street
Anytown, US 00000
555-555-5555

(Date)

(Name)
(Address)
(City, State, Zip)

Dear _____,

We appreciate the time you spent with us on _____ (date), and we wish to thank you for selecting our company to develop a design/build proposal to construct a _____ (description of facility) in _____ (location).

This letter, signed by both parties, will serve as authorization for our company to proceed with preliminary designs so that a budget for the project can be established. The cost of this initial design work will not exceed _____ (dollar amount). If this project advances to contract, this amount will be credited toward the total design fees.

We look forward to a successful relationship with your company and will proceed with the design work as soon as we receive the executed agreement from you.

Sincerely,

_____

(Name)

(Title)

_____

For (customer company name)

(Name, Title)

File Type: Word | **File Name:** 92-Design-Authorization.doc

Design/Build Cover Letter

ABC CONTRACTORS

123 Any Street
Anytown, US 00000
555-555-5555

(Date)

(Name)
(Address)
(City, State, Zip)

Dear _____,

We have enclosed our contract proposal regarding the construction of a new _____ (description of project) in _____ (location).

This proposal incorporates all of the building requirements that you have conveyed to our office. With your approval, the next step will be to proceed immediately to the completion of the construction drawings. We have incorporated into this proposal a schedule which shows the major steps and dates that will be necessary for completion of the project by _____ (completion date).

It is our intention, if you accept the terms and conditions of the proposal, to prepare a more formal contract to be approved on or about _____ (date). If you do not feel that this is practical, we suggest that you authorize us to continue to complete the construction drawings for the project at a cost not to exceed that shown in section _____ (section number) of our proposal.

We appreciate the opportunity you have given us to work with your company, and we look forward to the start of construction in the near future.

Sincerely,

_____

(Name)

(Title)

Design/Build Introduction Letter

ABC CONTRACTORS

123 Any Street
Anytown, US 00000
555-555-5555

(Date)

(Name)
(Address)
(City, State, Zip)

Dear _____,

We have been working on a bid for the project mentioned above, and have noticed some design possibilities that may be advantageous to you.

Whether or not these changes are to be incorporated into the forthcoming bid, I would like you to be aware of them. These modifications should save you a considerable amount of money, without affecting the quality of the installation.

Specifically, we think the following modifications would prove advantageous:

(describe changes and advantages here)

If you have any inclination to make such modifications, you may reach me at _____ (phone number) during normal business hours.

Sincerely,

_____

(Name)

(Title)

Design/Build Modification Letter

ABC CONTRACTORS

123 Any Street
Anytown, US 00000
555-555-5555

(Date)

(Name)
(Address)
(City, State, Zip)

Dear _____,

Enclosed you will find our bid on the _____ project according to the plans and specifications for this project which were prepared by _____ (name of architect and/or engineer). We also acknowledge the receipt of addendum _____, and have incorporated them into this bid amount.

In addition, in our extensive review of this project, we have noticed a number of areas where design changes may be advantageous to you. Specifically, they are as follows:

(detail changes and associated benefits here)

Thank you again for the opportunity of bidding on this project.

Sincerely,

_____

(Name)

(Title)

Design Estimate Letter

ABC CONTRACTORS

123 Any Street
Anytown, US 00000
555-555-5555

(Date)

(Name)
(Address)
(City, State, Zip)

Dear _____,

Pursuant to our conversations of the past few weeks, the following is an itemized preliminary estimate for the _____ (project name) in _____ (location).

The design of the _____ will be as follows:

(design detail)

The costs of this design are as follows:

ITEM _____ (costs)   COST _____ (dollars)

(add more lines as required)

TOTAL COST . . . . $ _____

This estimate was based upon plans dated _____, prepared by _____ (architect's name), and represents a complete, organized, and professional installation.

We look forward to working with you.

Sincerely,

_____

(Name)

(Title)

Submittal Letter

ABC CONTRACTORS

123 Any Street
Anytown, US 00000
555-555-5555

(Date)

(Name)
(Address)
(City, State, Zip)

Re: _____ (project name or designation)

Dear _____,

Enclosed you will find ___ sets of submittals for the _____ portion of the above referenced project.

We are expecting to receive ___ sets of reviewed submittals back in approximately _____(amount of time). If this is not correct, please notify us of when we can expect them.

As you know, time is critical since we cannot order materials until the submittals are approved.

To the best of our knowledge, these are all the materials you need. If we are not correct, please notify us immediately.

Thank you for your prompt service.

Sincerely,

_____

(Name)

(Title)

Hazmat Submittal Letter

ABC CONTRACTORS

123 Any Street
Anytown, US 00000
555-555-5555

(Date)

(Name)
(Address)
(City, State, Zip)

Dear _____,

Enclosed you will find the required sheets for hazardous materials that we will be using on the _____ project, according to EPA and OSHA regulations.

Although we believe these sheets to be in full compliance with the requirements, the regulations are somewhat confusing. If you find any deficiencies in these sheets, please notify us of the problems, and we will proceed with their correction.

Sincerely,

_____

(Name)

(Title)

General Delay Notice Letter

ABC CONTRACTORS

123 Any Street
Anytown, US 00000
555-555-5555

(Date)

(Name)
(Address)
(City, State, Zip)

Re: _____ (name of project)

Dear _____,

As you know, the project to which we are mutually committed is now several weeks behind schedule. Typically, one trade must complete work before another can begin, which means those scheduled near the completion of the project can be mistakenly blamed for a delay. In our case, that is simply not true. Throughout the course of this project we have remained on schedule, performing our work in a timely manner. We have not been responsible for any delays. As a protection and a way to explain our position, I have included a detailed account of the delays affecting the project at the moment, and the reason, in my opinion, for those delays.

Description:_____

_____

_____

At this point, my primary concern is that the project is scheduled for completion in _____ days, though _____ (name of trade) will not complete their work for at least _____ days, which means what time remains is not enough for us to complete our work. Please check the original project schedule; according to the schedule we should be working now. Because other trades have not completed their work, we are forced to wait. The purpose of this letter is to explain our position, and to ensure that the blame for any delays will not be assigned to us. Thank you for your consideration.

Sincerely,

_____

(Name)
(Title)

Delay Notice Letter – Other Trade

ABC CONTRACTORS

123 Any Street
Anytown, US 00000
555-555-5555

(Date)

(Name)
(Address)
(City, State, Zip)

Re: _____ (project name or designation)

Dear _____,

Within the last few weeks we have become increasingly concerned over delays in the progress of the project referenced above. Within the last few days, we realized the situation must be addressed; your failure to meet scheduling deadlines is a concern we can no longer ignore. As you know, work is often sequenced so that those responsible for a certain stage in the process must complete that stage before workers waiting to continue the process can proceed. Recently we were forced to lay off several workers because there was no work; these men depend on us to deliver work and subsequently a paycheck. We take our obligation to provide both very seriously. Contractors who fail to complete work as scheduled make it difficult if not impossible for us to honor our commitment to our workers and inevitably to our customers.

As a result of these delays, the completion date for this project, originally scheduled only _____ weeks from the date of this letter, is in serious jeopardy. Completing a project as per the original agreement is a priority with our company; failure to do so is a serious infraction of our commitment to our customer. We ask that you advise us immediately of your intentions regarding the progress of this project; we intend to deliver a completed product as per the original agreement, and we do not intend to assume any degree of liability for a problem for which we share no responsibility. Please contact us immediately to advise us of your intentions so that more serious consequences can be avoided.

Sincerely,

_____

(Name)

(Title)

cc: (associated parties)

Weather Delay Letter

ABC CONTRACTORS

123 Any Street
Anytown, US 00000
555-555-5555

(Date)

(Name)
(Address)
(City, State, Zip)

Re: _____ (project name or designation)

Dear _____,

In response to your letter of _____(date) regarding your concerns about delays in the work schedule for the project mentioned above, I offer as explanation the inclement weather conditions of the last few weeks, conditions which have seriously affected our ability to complete work as previously agreed.

As a result, we are requesting an extension of _____ days to the original contract completion date. Should you require documentation to support our request, we offer copies of our foreman's logs which include dates, weather conditions, and the specific jobs affected.

Unless we hear otherwise, we will proceed according to our revised schedule; however, we ask that you send written confirmation of your acceptance of our request.

Sincerely,

_____

(Name)

(Title)

Job Problem Letter

ABC CONTRACTORS

123 Any Street
Anytown, US 00000
555-555-5555

From: _____

To: _____ (foreman, superintendent, etc.)

Re: _____ (project name)

_____: (name)

I'm afraid we've received some complaints about your project. Specifically, they are:

(detail complaints)

Please let me know if these complaints are accurate, and then let's decide what we are going to do about them. We cannot wait on this. It must be resolved quickly so it doesn't hurt the project.

Please be sure that you have detailed answers to me by _____ (date or time).

Thank you,

_____

(sign name or initial here)

Subcontractor Problem Letter

ABC CONTRACTORS

123 Any Street
Anytown, US 00000
555-555-5555

(Date)

(Name)
(Address)
(City, State, Zip)

Re: _____ (project name or designation)

Dear _____ ,

As you know, the payment scheduled for _____ (date) has not been received in our offices. While I realize your ability to meet your obligations is dependent on payment from the owner, I must remind you that our company is small and has limited reserve funds available to cover problems; certainly we cannot continue to work if our reserves are exhausted. I also wish to remind you that the original contract addressed payment schedules and amounts. With that in mind I ask you to meet with the owner immediately to address this situation; please keep us apprised of your intentions and progress.

In closing, if we do not receive payment by _____ (date), we will be forced to withdraw from the project and seek legal advice for collecting the money owed to us. I am sure the expense involved in acquiring a new subcontractor, not to mention the additional delays that will result, will more than justify a quick resolution of this dilemma.

I will be waiting to hear from you.

Sincerely,

_____

(Name)

(Title)

cc: (parties involved)

Pull Off Letter #1

ABC CONTRACTORS

123 Any Street
Anytown, US 00000
555-555-5555

(Date)

(Name)
(Address)
(City, State, Zip)

Re: _____ (project name or designation)

Dear _____,

This letter serves as final notice; if the conditions outlined in my letters of _____ (date) and _____ (date) are not addressed within five days, we will end our involvement in this project. Should such action become necessary, please keep in mind that we will require payment of expenses for dismantling our equipment and crews as well as payment for reassembling the two if and when we return to the project. These expenses are included in the list of possible charges outlined in the original contract.

While I regret that such steps may become necessary, we feel we have no option; present conditions demand that we take steps to protect our interests.

Sincerely,

_____

(Name)

(Title)

Pull Off Letter #2

ABC CONTRACTORS

123 Any Street
Anytown, US 00000
555-555-5555

(Date)

(Name)
(Address)
(City, State, Zip)

Re: _____ (project name or designation)

Dear _____ ,

Please be advised that because of the problems I outlined to you in my letters of _____ and _____ (dates), we have found it necessary to pull off of this project.

Please be further advised that we will require payment for both demobilization and mobilization expense before we can return to the project. This was stated explicitly in my letter of _____ (date). You can find itemizations of these expenses on the schedule of values that we submitted to you prior to commencing the project.

Also be advised that we are seeking legal counsel on this matter, and that further actions regarding it may follow.

I am sorry that drastic steps are necessary, but we can not allow ourselves to be put at risk.

Sincerely,

_____

(Name)

(Title)

Notice of Completion Letter

ABC CONTRACTORS

123 Any Street
Anytown, US 00000
555-555-5555

(Date)

(Name)
(Address)
(City, State, Zip)

Dear _____,

Please be aware that our work at the following location has been completed:

PROJECT NAME: _____

ADDRESS: _____

PHONE: _____

CONTRACT AMOUNT: _____

Note that we have enclosed our final waiver, our final invoice, and your guarantee.

Sincerely,

_____

(Name)

(Title)

Change Order Letter

ABC CONTRACTORS

123 Any Street
Anytown, US 00000
555-555-5555

(Date)

(Name)
(Address)
(City, State, Zip)

Re: _____ (project name or designation)

Dear _____,

This letter is to confirm that we have been requested to perform work in addition to our contract for the project noted above.

Specifically, _____ (person's name) of your company instructed us to_____ (extra work being done), which he/she understood to be additional to our contract amount.

We are proceeding with this work, per _____'s (name) instructions on _____ (date).

If there will be any special billing instructions for this work, please let us know. If not, we will simply invoice you as we normally do for small jobs.

Thank you for the additional work.

Sincerely,

_____

(Name)

(Title)

CHAPTER FOUR

Contracts

This chapter contains useful forms and contracts based on your business correspondence with your customers.

FORMS

Proposal and Contract

This is a very simple proposal form. It can be used for any type of construction project and is especially well suited to small projects. Once accepted by the customer, it is a legal contract.

Subcontract Rider #1

This contract rider is written for the purpose of protecting the subcontractor's interests in the face of contracts and contract documents that were not written with his/her best interests at heart. Feel free to add to this rider or remove from it, as seems appropriate.

Subcontract Rider #2

This is another subcontractor's rider. Again, feel free to modify as required.

Project Management Agreement

This is an agreement between a project management company and a project owner. The terms it contains are fairly standard, but you may wish to modify them for each project.

Deficiency Notice

Use this form as a notice to a subcontractor that there are required contractual items missing.

Contract Transmittal

This is a cover letter for the delivery of a contract.

Contract Statement

Use this form to update contract amounts. This would be used for a monthly update, consolidating many smaller changes into a single statement.

Subcontract Insurance Verification Log

Especially useful to general and prime contractor, this form is for making sure that the subcontractors have the required insurance and that you have the necessary proof of it.

Proposal and Contract

Date: _____

To: _____

_____

_____

Project Location: _____

_____

WE PROPOSE TO FURNISH ALL MATERIALS, EQUIPMENT, AND SUPERVISION TO COMPLETE THE FOLLOWING:

_____

_____

_____

_____

_____

_____

_____

_____

_____

_____

TERMS: _____

_____

Very Truly Yours,

_____

ACCEPTED:

Please proceed with the above work.

Signature: _____

Date: _____

Subcontract Rider #1

Project: _____

This rider is attached to, and made a part of, the contract dated _____ (date) between _____ (general contractor), hereafter called "contractor," and _____ (subcontractor), hereafter called "subcontractor."

Acceptance of this proposal by Contractor shall be construed as acceptance of all terms and conditions recited herein which shall supersede any conflicting term in any other contract document. Any of the Contractor's terms and conditions in addition or different from this proposal are objected to and shall have no effect. Contractor's agreement herewith shall be evidenced by Contractor's signature hereon or by permitting Subcontractor to commence work for project.

1. Subcontractor shall be paid monthly progress payments on or before the 15th of each month for the value of work completed plus the amount of materials and equipment suitably stored on or off site. Final payment shall be due 30 days after the work described in the Proposal is substantially completed. No provision of this agreement shall serve to void the Subcontractor's entitlement to payment for properly performed work or suitably stored materials or to require the Subcontractor to continue performance if timely payments are not made to Subcontractor for suitably performed work or stored materials or to void Subcontractor's right to file a lien or claim on its behalf in the event that any payment to Subcontractor is not made in a timely manner.

2. The Contractor will withhold no more retention from the Subcontractor than is being withheld by the Owner from the Contractor with respect to the Subcontractor's work.

3. All sums not paid when due shall bear an interest rate of $1\frac{1}{2}\%$ per month or the maximum legal rate permitted by law, whichever is less; and all costs of collection, including a reasonable attorney's fee, will be paid by Contractor.

4. No back-charges or claims of the Contractor for services shall be valid except by an agreement in writing by the Subcontractor before the work is executed, except in the case of the Subcontractor's failure to meet any requirement of the subcontract agreement. In such event, the Contractor shall notify the Subcontractor of such default, in writing, and allow the Subcontractor reasonable time to correct any deficiency before incurring any cost chargeable to the Subcontractor.

5. Contractor is to prepare all work areas so as to be acceptable for Subcontractor work under the subcontract. Subcontractor will not be called upon to start work until sufficient areas are ready to ensure continued work. The Contractor shall furnish all temporary site facilities including suitable storage space, hoisting, and temporary electrical and water at no cost to Subcontractor.

6. Subcontractor shall be given a reasonable time in which to make delivery of materials and/or labor to commence and complete the performance of the contract. Subcontractor shall not be responsible for delays or defaults where occasioned by any causes of any kind and extent beyond its control, including, but not limited to: delays caused by the owner, general contractor, architect and/or engineers, delays in transportation, shortage of raw materials, civil disorders, labor difficulties, vendor allocations, fires, floods, accidents, and acts of God. Subcontractor shall be entitled to equitable

Subcontract Rider #1 (*cont.*)

adjustment in the subcontract amount for additional costs due to unanticipated project delays or accelerations caused by others whose acts are not the Subcontractor's responsibility and to time extensions for unavoidable delays. The Contractor shall make no demand for liquidated damages for delays in excess of the amount specified in the subcontract agreement and no liquidated damages may be assessed against Subcontractor for more than the amount paid by the Contractor for unexcused delays to the extent actually caused by Subcontractor.

7. The Subcontractor's equipment and work are guaranteed for a period of one year from the date of substantial completion or use by the Contractor or the Contractor's customer, whichever is earlier. THIS WARRANTY IS IN LIEU OF ALL OTHER WARRANTIES, EXPRESS OR IMPLIED, INCLUDING ANY WARRANTIES OF MERCHANT OR FITNESS FOR A PARTICULAR PURPOSE. The exclusive remedy shall be that Subcontractor will replace or repair any part of his/her work which is found to be defective. Subcontractor shall not be responsible for damage or defect caused by abuse, modifications not executed by the Subcontractor, improper or insufficient maintenance, improper operation, or normal wear, tear, and usage.

8. Work called for herein is to be performed during Subcontractor's regular working hours. All work performed outside of such hours shall be charged for at rates or amounts agreed upon by the parties at the time overtime is authorized.

9. Contractor shall, if the Owner does not, purchase and maintain all risk insurance upon full value of the entire work and/or materials delivered to the jobsite, which shall include the interest of Subcontractor.

10. The Subcontractor shall indemnify and hold harmless the Contractor, Owner, Architect, or others from damages only to the extent such damages were caused by any negligent act or omission of the Subcontractor or anyone for whose acts the Subcontractor is liable.

11. The subcontract form used between the Subcontractor and the Contractor will be AIA Standard Form Subcontract Document A401. Where there is a conflict between provisions of either the AIA Standard Form or the contract documents between the Owner and Contractor and this Proposal, then this Proposal shall govern.

Date: _____

Contractor: _____   Subcontractor: _____

By: _____   By: _____

Title: _____   Title: _____

File Type: Word | **File Name:** 109-Subcontract-Rider-1.doc

Subcontract Rider #2

Date: _____

Project Designation: _____

This rider is attached to, and made a part of, the contract dated _____ (date) between _____ (general contractor), hereafter called "contractor," and _____ (subcontractor), hereafter called "subcontractor."

The parties to this Subcontractor Policy agree as follows:

1. Upon acceptance of a bid by the contractor, the subcontractor shall become an integral component of the project. The subcontractor and contractor must work together to ensure success. Should the subcontractor anticipate any problems, find any code violations, or find technical problems, it is the duty of the subcontractor to report this immediately to the contractor.

2. All bids shall be based upon the subcontractor's site visit. Therefore, the subcontractor will be held accountable for any conditions at the job site that should have been discovered upon investigation.

3. Upon acceptance of the subcontractor's bid, the subcontractor and his/her employees shall conduct themselves in a professional and workmanlike manner. The reputation of both the contractor and the subcontractor are directly impacted by the conduct of all involved with the project.

4. If any of the contractor's clients request any work directly from the subcontractor, the subcontractor must obtain written permission to perform such work from the contractor. In addition, if the subcontractor is approached on the job site by a potential new client, the subcontractor must refer such potential new client to the contractor's office. Any subcontractor who takes such a job directly shall be in direct violation of this agreement.

5. The subcontractor may not make any changes or modifications to the work being performed without the prior written consent of the contractor. If conditions require a verbal permission, the subcontractor must immediately thereafter file a written change order with the contractor. The subcontractor is prohibited from providing prices directly to the client.

6. Prior to commencing any work, the subcontractor must provide the contractor with a copy of his/her general liability or worker's compensation coverage. If the subcontractor fails to comply with this requirement prior to any and all payouts, the contractor shall deduct _____% from the amount due to the subcontractor for the general liability and _____% for the workers' compensation.

7. The subcontractor must immediately inform the contractor of any issues that may delay the work of the subcontractor. If the subcontractor determines, while on site, that additional work is necessary before the subcontractor can proceed, the contractor shall be promptly informed of this situation.

8. All project related inquiries shall be referred to the contractor.

For Contractor: For Subcontractor:

_____   _____

(Name)   (Name)

(Title)   (Title)

Project Management Agreement

Project Designation: _____

Address: _____

                            _____

                            _____

This agreement is made this _____ (number) day of _____ (month), 20___ (year) between _____ (owner), hereinafter called the Owner, whose address is _____ (address), and _____ (project manager), hereinafter called the Project Manager whose address is _____ (address).

The parties above hereby agree as follows:

The Project Manager agrees to use his best efforts in a professional and workmanlike manner to manage and see to completion the construction of a single family residence (herein called the "Project") for _____ upon the real property located at _____ (address).

The Project Manager's Roles and Responsibilities shall be as follows:

PRE-CONSTRUCTION PHASE

1. Assist in finding a lender.

2. Obtain necessary permits and handle interfaces with the building department (general, grading, electrical, mechanical, plumbing, etc.).

3. Provide a material takeoff for lumber and framing hardware.

4. Manage the bidding process, obtaining at least three bids for each major trade.

5. Act as the clearing-house for information to bidding subcontractors.

6. Check the references of each subcontractor.

7. Assemble presentation to the lender.

8. Provide a detailed cost breakdown sheet for the lender.

9. Develop a critical path time-line with receipt of Certificate of Occupancy targeted for _____ (date).

10. Act as the contractor of record with the lender.

11. Recommend to the Owner the most qualified and reasonably priced subcontractor for each phase of construction based on the bids submitted.

12. Confirm and finalize the contract documents with each subcontractor.

13. Develop a list of contract addendum covering areas of payment disbursement, safety, workers compensation, cleanup, and craftsmanship standards.

Project Management Agreement (*cont.*)

CONSTRUCTION PHASE

The Project Manager shall ensure that the Project is constructed and completed in strict conformance with the plans and specifications and that laws, ordinances, rules, and regulations of the applicable governmental authorities are adhered to. Further, the Project manager agrees to:

1. Be the Owner's representative in the field.

2. Meet with Owner on a regular basis to review progress.

3. Coordinate utility connections.

4. Coordinate with geologists and/or soil engineers.

5. Coordinate with the grader.

6. Supervise day-to-day construction, ensuring that the Project is built as intended by the Owner and the Designers and per plans and specifications.

7. Ensure that neighboring properties and owners are protected and respected.

8. Ensure that the subcontractors maintain the highest degree of craftsmanship.

9. Schedule subcontractors and suppliers.

10. Review the Workers Compensation policy of each subcontractor and have a Certificate of Workers Compensation sent to the Project Manager by its insurance carrier.

11. Manage incidental day labor.

12. Keep track of change orders.

13. Verify materials deliveries to ensure they are in accordance with purchase orders.

14. Authorize payment of subcontractors based upon ongoing review of their work.

15. Obtain lien releases.

16. Provide on-site problem solving with and between subcontractors.

17. Coordinate with interior designer.

18. Call for inspections and coordinate with Building Department.

19. Ensure the job site is cleaned on a regular basis. (labor costs to be met from Owner's funds)

20. Ensure that all workers maintain the highest safety standards.

21. Pass along to the Owner any discounts the Project Manager is entitled to as a General Contractor without charging any markups.

Project Management Agreement (*cont.*)

OWNER'S ROLE AND RESPONSIBILITIES

1. Finalize negotiations with a lender and secure the construction loan.

2. Apply for and pay for the General Building Permit. (All other permits are to be paid for and taken out by each respective subcontractor.)

3. All construction expenses are to be borne by the Owner. The Project Manager is not to be held liable for any unpaid bills.

4. Sign contracts with each subcontractor upon the recommendation of the Project Manager.

5. Clear any work or changes on the Project through the Project Manager.

6. Maintain a petty cash account for incidental purchases. (Project Manager shall keep track of these expenses.)

7. Provide a job site phone upon commencement of rough grading, the cost of which is to be included in the construction loan.

8. Provide a temporary construction field office available to the Project Manager upon commencement of rough grading (cost to be included in the construction loan).

9. Accept liability for the theft or destruction of any building materials.

10. Provide all appropriate insurance policies (public liability, course of construction, fire, theft, XCU, etc.).

11. Provide a chain-link fence around the Project area.

12. File a Notice of Completion within 5 days of substantial completion of the Project.

FEE SCHEDULE

In consideration for the above services rendered by the Project Manager, the Owner agrees to pay the Project Manager a Base Fee of 10% of the cost of construction, based on the amount of the construction loan plus, as an incentive to bring the Project in under budget, a Final Draw of 50% of any unused loan funds that exist when the Notice of Completion is filed by the Owner and construction budgeted for in the construction loan is completed.

The Base Fee is to be computed based on the total hard construction costs approved in the construction loan including permit costs, appliances, and fixtures. Any soft costs included in the construction loan such as design, engineering, interior design, as well as furniture and drapery costs, are not to be considered in determining the Base Fee.

The Base Fee is to be broken down into installments as follows:

- An initial retainer of $_____ (amount) to be paid by _____ (date), to cover the first month of pre-construction services.

- A monthly installment of _____% of the base fee to be paid on the 1st of each following month up to _____ (date or time).

Project Management Agreement (*cont.*)

The final draw is to be paid within 15 days of filing the Notice of Completion, based on the following:

- As a mutually-beneficial incentive to finish the construction in a timely manner and to meet the completion goal stated above, the Project Manager shall receive the full 50% final draw as stipulated above if construction of the living space is completed to warrant a Certificate of Occupancy by _____ (date).

- If Certificate of Occupancy is not granted by _____ (date), 10% shall be deducted from the Project Manager's final draw for each 15 calendar day increment beyond _____ (date).

- If progress is delayed for any reason beyond the Project Manager's control, the time frame for computing the final draw shall be shifted beyond _____ (date) by the amount of the delay. Examples of such delays include: prolonged adverse weather, earthquakes, fire, work stoppages, and approval delays.

- Cost overruns, in any particular construction budget category or allowance, made by the Owner during construction that are a result of changes or upgrades not called out in the plans and specifications, are not to be considered as part of any remaining construction funds for purposes of determining the final draw.

We hereby execute this Agreement at _____ (location) on _____ (date).

For Owner: For Project Manager:

_____   _____

(Name)   (Name)

(Title)   (Title)

Deficiency Notice

ABC CONTRACTORS

123 Any Street
Anytown, US 00000
555-555-5555

Date: _____ From: _____

To: _____ Project no.: _____

The following items have not been submitted to us. Return by: _____

[ ] Contract   [ ] Warranty   [ ] Minority Compliance Form

[ ] Purchase Order   [ ] Shop Drawings   [ ] Product Literature

[ ] Certificate of Insurance   [ ] As-built Drawings   [ ] Other:_____

[ ] Sub-schedule   [ ] Samples

[ ] Submittals   [ ] Colors/Textures

[ ] Cost Breakdown   [ ] Materials List

Please forward this material immediately for our review.

Note:

_____
_____
_____
_____
_____
_____
_____
_____
_____
_____
_____
_____

Contract Transmittal

ABC CONTRACTORS

123 Any Street
Anytown, US 00000
555-555-5555

Date: _____   Project No.: _____

Contractor: _____   Project Name: _____

Address:

_____

_____

_____

Dear _____:

Enclosed please find two copies of our contract for the project referenced above.

Please sign both copies and return them to this office with your certificate of insurance. A countersigned copy will be forwarded to you upon our receipt of your signed copies.

Please note your project name and number for future reference. Please make reference to this project number in all of your correspondence with us.

Thank you.

Sincerely,

_____

Enclosures

Contract Statement

ABC CONTRACTORS

123 Any Street
Anytown, US 00000
555-555-5555

Date: _____   Address: _____

Project Name: _____   _____

Owner: _____   _____

Change order record:

Change orders to date   _____

Change order increase   _____

Change order decrease   _____

Net change   _____

Statement of contract:

Original contract amount   _____

Net change orders (see above)   _____

Adjusted contract amount   _____

Billing history:

Adjusted contract amount, billed to date   _____

Less _____% retainage   _____

Total adjusted contract less retainage   _____

Current invoice amount   _____

Less _____% retainage   _____

Current invoice amount less retainage   _____

                          PAY THIS AMOUNT   [   ]

Certified by:

The undersigned certifies that to the best of his/her knowledge the work covered by this statement has been completed in accordance with the contract documents, that previous invoices were issue, and the current payment shown is now due.

_____   Date: _____

Subcontractor Insurance Verification Log

ABC CONTRACTORS

123 Any Street
Anytown, US 00000
555-555-5555

Address: _____   Date: _____

_____   Project No.: _____

Phone: _____   Project Name: _____

Subcontractor	Workers' Compensation	General liability	Federal ID #
	[ ] Expires:_____ Certificate #: _____ Ins. carrier:	[ ] Expires:_____ Certificate #: _____ Ins. carrier:	
	[ ] Expires:_____ Certificate #: _____ Ins. carrier:	[ ] Expires:_____ Certificate #: _____ Ins. carrier:	
	[ ] Expires:_____ Certificate #: _____ Ins. carrier:	[ ] Expires:_____ Certificate #: _____ Ins. carrier:	
	[ ] Expires:_____ Certificate #: _____ Ins. carrier:	[ ] Expires:_____ Certificate #: _____ Ins. carrier:	
	[ ] Expires:_____ Certificate #: _____ Ins. carrier:	[ ] Expires:_____ Certificate #: _____ Ins. carrier:	
	[ ] Expires:_____ Certificate #: _____ Ins. carrier:	[ ] Expires:_____ Certificate #: _____ Ins. carrier:	
	[ ] Expires:_____ Certificate #: _____ Ins. carrier:	[ ] Expires:_____ Certificate #: _____ Ins. carrier:	
	[ ] Expires:_____ Certificate #: _____ Ins. carrier:	[ ] Expires:_____ Certificate #: _____ Ins. carrier:	

CHAPTER FIVE

Forms and Letters for Your Vendors

This chapter contains useful forms and contracts based on your business correspondence to your customers.

FORMS

Purchase Order

This is a basic purchase order form. Notice that the form specifies a three-part number to be included on two copies of an invoice. If you use these numbers carefully, they will be a great aid to you in the tracking and checking of invoices and billings. Note also that this purchase order is designed to be signed by the vendor that receives it. This verifies that the vendor will have to comply with the explicit terms of the purchase order and any attached cover letter. Doing this has helped many contractors avoid problems.

Contractor Qualification Form

You will find this to be a very thorough qualification form. By the time you finish reviewing this form and the financial statement and insurance certificate that are also requested, you will have an extremely detailed portrait of the prospective subcontractors. Once this form is complete, you can call the phone numbers that are listed on it (bonding company, bank, insurance agent) to verify that the information given to you is correct. You should also call the references given on the form to verify the subcontractor's abilities and service characteristics.

LETTERS

Lost-Bid Letter

This is a polite notice to a bidder that he/she did not win the job.

Request for Credit

This is a simple letter requesting a charge account from a vendor. This letter presupposes that you have been buying from the vendor for some time on a C.O.D. basis. If your company has a good credit record, you can simply state that you would like a charge account, state your credit references, and inform the vendor that they can call you for further credit information.

Bad Invoice Letter

This letter is written in response to an incorrect invoice being received from the vendor. Specifically, the problem is that the invoices do not match with the purchase order that authorized them. These problems are sometimes the result of accidents, but sometimes seem almost on purpose. The letter details the problem and nicely says that further payment will not be made until the invoices are corrected, showing the originally agreed-upon prices.

Invoice Problem Letter

This is a letter to a vendor that has been making mistakes on his or her billings. In this letter you should describe the problems in detail, if you have not previously. This is a polite letter addressing a problem that is neither purposeful or severe.

Paying Late Letter

This letter is a request by the contractor for an extension of credit terms due to a financial shortfall. Because of a financial difficulty, the contractor is being forthright and asking the vendor for an extension of time. To do this properly, you must inform the vendor as soon as you know that there will be a delay, briefly explain the cause of the delay, and specify a date on which the vendor can expect payment.

Cancellation Letter

In some situations, you may be forced to cancel orders from vendors that are not delivering their products as agreed. In such situations you will need to send them a legal cancellation of the order. This letter should be sent to the delinquent vendor via certified return-receipt mail. (Or via an overnight service that will give you a signed receipt of delivery.) As long as you detail the reasons for cancellation and have proof that the letter was received, you are legally protected. The tone of this letter should be sympathetic, but firm.

Hazmat Request Letter

This is a request to a vendor/subcontractor for hazmat information sheets. You should be able to get hazmat sheets on any hazardous materials you buy from the vendor that sells it. The letter states that the vendor should have the sheets sent to you quickly since the government's requirements upon you are strict. You can also write a section in your purchase orders requiring hazmat sheets on all applicable materials within a certain number of days.

Material Delivery Letter

This is a letter that outlines the delivery requirements that you will require from a vendor. The letter references your purchase order for the material and clearly defines the delivery requirements. Note that a 24-hour prior notification is required of the vendor. This is something that you should use as a standard procedure since it can help you avoid a lot of last minute hassles, especially for unloading difficult items. The 24-hour prior notification should also be written into your purchase order, with this letter being used as a double verification.

Subcontractor Payment Problem Letter

This is essentially a letter requesting prompt payment. It can be used by almost any company in the construction business. It is written from the most common standpoint, that of the subcontractor requesting payment from a general contractor. For all other situations the letter will require modification. The letter explains exactly how long your company has been waiting for its payment and why you are not capable of continued waiting. Notice that this letter assumes that the only reason the general contractor is not paying you is because he or she has not yet been paid. The letter goes on to explain that neither you nor the general contractor are required by contract to wait so long, and that you need them to approach the owner about getting the situation straightened out.

The second-to-last paragraph of this letter is the "playing hardball" paragraph, which you may wish to remove on a first request. It mentions that you are sending a copy of this letter to the owner's representative. Assuming that the general contractor is not being honest, and is in fact holding your money, this paragraph is likely to secure them some grief from the owner. (Which, in such a situation, they would richly deserve.) For your first letter, you may wish to simply threaten such action, by revising this paragraph to read:

"If you are unable to resolve this situation by the end of the week, we will be forced to send a copy of this letter to the owner. We have been advised that we must do this to assure our legal protection."

Subcontract Cancellation Letter

This letter is a final resort to a problem with a subcontractor or sub-subcontractor. By the time you write this letter, you will have made the decision that the original subcontractor (the one who is getting this letter) is not able to live up to his or her contract. In this situation, you have two objectives:

1. To establish your legal standing in the termination of the subcontractor agreement so that you will have no future liability.

2. To make sure that you can get the subcontract completed with no additional expense (no cost overruns).

If you send this letter with no previous letters, your legal situation will not be solid; this is the last of a line of letters, not the first.

This letter specifies that you will be back-charging the recipient for any cost overruns on the completion of their contract. Assuming that you have a well-written subcontract agreement with them, this can be legally done. But bear in mind that the subcontractor may file for bankruptcy rather than pay you for a cost overrun. Because of this, do your best (even before you write this letter) to assure that you can find another subcontractor who can do the job within your budget. Send this letter via certified, return-receipt mail only.

Purchase Order Letter

This is a cover letter to a purchase order. It reiterates that everything included in the order must be delivered exactly as specified, unless the changes are agreed to in writing. To get an extra measure of legal protection you should include a cover letter like this one with every purchase order that you issue.

Purchase Order

To: _____   Date: _____

Ship to: _____

Shipping instructions: _____

_____

DELIVER THE FOLLOWING - - - - - - - - - - - - - - - - AS SPECIFIED HEREIN

_____

Submit two invoices bearing this P.O. number:

_____ _____ _____   By: _____

Received and acknowledged by: _____

Date: _____

Contractor Qualification Form

Date: _____

INSTRUCTIONS: You may elaborate on the requested information, or add to it if you think it will be useful for evaluation of your firm's capabilities. Simply attach extra information to this form. The information we receive will be used by _____ (your company's name) as a basis for determining bid sources. PLEASE ATTACH TO THIS FORM a copy of your certificate of insurance indicating current limits and a current financial statement.

Firm: _____

Address: _____

_____

Federal ID#: _____

Bank reference: _____

Address: _____

Phone: _____

Bonding capacity: _____

Work currently bonded: _____

Bonding Co.: _____

Bonding Co. Best rating: _____

Bonding agent phone: _____

Ins. Agent: _____

Ins. agent phone: _____

Phone #: _____

Fax #: _____

President: _____

Contact: _____

Date Co. began: _____

Years in trade: _____

Value of equipment: _____

Stockholder equity: _____

Percent work done by own workers: _____

Work under contract: _____

Volume, last year: _____

Average sales, last 3 yrs.: _____

Merit shop? Yes ___ No ___

Union affiliation? _____

Comply with EEO requirements? Yes ___ No ___

Staff you employ:

Engineers ____ CPM Schedulers ____ Project managers ____

Purchasing agents ____ Estimators ____ Draftsmen ____

In-house engineering capacity? Yes ___ No ___

In-house fabrication capacity? Yes ___ No ___

Fabrication area square feet _____

Do you have a safety program? Yes ___ No ___

Have you been cited for serious OSHA violations? Yes ___ No ___

If Yes, please explain circumstances separately.

File Type: Word | **File Name:** 117-ContractorQualification.doc

Contractor Qualification Form (*cont.*)

Qualified as MBE or WBE? Details: _____

_____ Attach certificates.

Geographic areas where you are interested in bidding work:

_____

Type of work you wish to bid: _____

_____

REFERENCES

Non-government work done in the past three years:

Customer: _____

Phone #: _____

Contract amount: _____

Description of project: _____

_____

_____

Completion: On-time _____ Ahead of schedule _____ Late _____

Customer: _____

Phone #: _____

Contract amount: _____

Description of project: _____

_____

_____

Completion: On time _____ Ahead of schedule _____ Late _____

Customer: _____

Phone #: _____

Contract amount: _____

Description of project: _____

_____

Completion: On time _____ Ahead of schedule _____ Late _____

Contractor Qualification Form (*cont.*)

Government work completed in the past three years:

Customer: _____

Phone # _____

Contract amount: _____

Description of project: _____

_____

_____

Completion: On time _____ Ahead of schedule _____ Late _____

Customer: _____

Phone # _____

Contract amount: _____

Description of project: _____

_____

_____

Completion: On time _____ Ahead of schedule _____ Late _____

Customer: _____

Phone # _____

Contract amount: _____

Description of project: _____

_____

_____

Completion: On time _____ Ahead of schedule _____ Late _____

Lost-Bid Letter

ABC CONTRACTORS

123 Any Street
Anytown, US 00000
555-555-5555

(Date)

(Name)
(Address)
(City, State, Zip)

Re: _____ (name of project)

Dear _____,

We have evaluated all the bids submitted to us on the project noted above, and we have decided to award this contract to another bidder.

We appreciate the interest you have shown in this project and the time and effort you spent developing your proposal.

We look forward to the possibility of working with you in the near future.

Sincerely,

_____

(Name)

(Title)

Request for Credit Letter

ABC CONTRACTORS

123 Any Street
Anytown, US 00000
555-555-5555

(Date)

(Name)
(Address)
(City, State, Zip)

Dear _____ ,

We have been purchasing merchandise from you for _____ (period of time) C.O. D. Now we would like the convenience of a charge account.

Please let us know what information and references are necessary as well as your credit terms.

Sincerely,

_____

(Name)

(Title)

Bad Invoice Letter

ABC CONTRACTORS

123 Any Street
Anytown, US 00000
555-555-5555

(Date)

(Name)
(Address)
(City, State, Zip)

Dear _____,

A discrepancy has arisen regarding our Purchase Order #_____, and your invoices for this material. I refer specifically to invoices #___, ___, ___, ___, and ___.

If you will compare your copy of our purchase order to these invoices, you will notice the variance between the quoted and actual prices.

Since we are not aware of any agreement to change the prices, I assume that this was an error on your company's part. Please send corrected invoices as soon as possible so that we may process them and get payment to you without further delay.

Sincerely,

_____

(Name)

(Title)

File Type: Word　|　**File Name:** 120-Bad-Invoice-Ltr.doc

Invoice Problem Letter

ABC CONTRACTORS

123 Any Street
Anytown, US 00000
555-555-5555

(Date)

(Name)
(Address)
(City, State, Zip)

Re: _____ (project name or designation)

Dear _____,

Recently we have noticed a significant number of invoices from you with errors. While we understand the occasional error, we do not understand the record number witnessed recently, nor do we feel it is our duty to spend company time auditing and correcting invoices. To the contrary, that is a duty belonging to your accounting department. Please address this situation immediately to ensure that future invoices are error free.

Thank you,

_____

(Name)

(Title)

Paying Late Letter

ABC CONTRACTORS

123 Any Street
Anytown, US 00000
555-555-5555

(Date)

(Name)
(Address)
(City, State, Zip)

Dear _____,

I am writing to formally request an extension for payment of the bill in the amount of $_____. Because a check from a client on which we were dependent for making payment to you has been delayed until _____, we, too, must ask for additional time. We sincerely hope this request does not inconvenience you. Please advise if you feel you cannot comply and we will seek some other solution.

Sincerely,

_____

(Name)

(Title)

Cancellation Letter

ABC CONTRACTORS

123 Any Street
Anytown, US 00000
555-555-5555

(Date)

(Name)
(Address)
(City, State, Zip)

Re: _____ (project name or designation)

Dear _____,

Since you are still not able to send us the material listed in our purchase order number _____, we are forced to cancel this order. As you know, we needed these materials for the _____ project, and they were needed by a very specific date as shown on our original purchase order. We are put in jeopardy by these delays and cannot wait any longer.

I am sorry that such a step is necessary, but we really have no other choice.

Sincerely,

_____

(Name)

(Title)

Hazmat Request Letter

ABC CONTRACTORS

123 Any Street
Anytown, US 00000
555-555-5555

(Date)

(Name)
(Address)
(City, State, Zip)

Re: _____ (project name or designation)

Dear _____,

Please forward to us the hazardous material sheets that are specified in our contracts.

Please remember also that these sheets should be filled out according to all EPA and OSHA requirements. You should keep copies for your own records and forward duplicates to us for our records.

Please attend to this matter with diligence, as the government's requirements are very specific.

Sincerely,

_____

(Name)

(Title)

Material Delivery Letter

ABC CONTRACTORS

123 Any Street
Anytown, US 00000
555-555-5555

(Date)

(Name)
(Address)
(City, State, Zip)

Re: _____ (project name or designation)

Dear _____,

Please be advised that we need the _____ materials for this project delivered on
_____ (date). Delivery should be made directly to the job site at _____
(address).

As noted in our purchase order #_____, this material should be delivered during normal business
hours, and we should be notified 24 hours prior to delivery.

Thank you in advance for your compliance with the purchase order requirements.

Sincerely,

_____

(Name)

(Title)

Subcontractor Payment Problem Letter

ABC CONTRACTORS

123 Any Street
Anytown, US 00000
555-555-5555

(Date)

(Name)
(Address)
(City, State, Zip)

Re: _____ (project name or designation)

Dear _____,

As you know, we have not yet received our _____ (date of payment period) payment on this project. This is a serious problem to us as we are a small company and we do not have immense cash reserves to carry us through a long waiting period.

I understand that you are waiting for money from the owner so you can pay us, but we are being put in a difficult spot. The terms of our contract with you (and yours with the owner) call for prompt payment. This does not appear to be happening, and it is straining us.

At this point, we very much need you to approach the owner about this problem. If we do not receive payment very soon, we will have to pull off of the project, and seek legal council regarding our financial protection. I know that this may seem rather extreme, but it is our only recourse to simply waiting.

As you will find noted at the bottom of this letter, we are sending a copy to the owner's representative, _____ (name).

Please let us know how you are progressing on this matter within the next 5 days. Thank you for your help.

Sincerely,

_____
(Name)

(Title)

Subcontract Cancellation Letter

ABC CONTRACTORS

123 Any Street
Anytown, US 00000
555-555-5555

(Date)

(Name)
(Address)
(City, State, Zip)

Re: _____ (project name or designation)

Dear _____,

Because you have failed to comply with our requests as stated in our letter dated _____, we are forced to terminate your subcontract agreement. As we stated in our letter, we cannot afford to put our project or our customer's trust at risk; we simply must ensure the success of this project in a timely manner.

We are currently canvassing the job market for a replacement to complete the work you began; while we will make every effort to ensure the amount of this new contract is fair to you and your replacement, we must charge this expense against the amount due you under the original agreement. If these charges exceed what we owe you, we will bill _____ (subcontractor's company's name) for the balance.

We regret that this action became necessary, but after carefully considering our options, we concluded there is no other solution.

Sincerely,

_____

(Name)

(Title)

Purchase Order Letter

ABC CONTRACTORS

123 Any Street
Anytown, US 00000
555-555-5555

(Date)

(Name)
(Address)
(City, State, Zip)

Dear _____,

Enclosed you will find our Purchase Order #_____, for _____ (type of equipment, materials, or service) on the _____ project. Please deliver the goods specified at the times specified and at the price noted on the purchase order.

Please remember that everything must be done exactly as shown on the order. No deviations will be allowed without our written consent.

Please pay careful attention to the notes regarding time and method of delivery, specifically that we want 24 hour notification prior to taking delivery.

Thank you for your service and attention to these details.

Sincerely,

_____

(Name)

(Title)

CHAPTER SIX

Legal Contracts, Agreements, and Letters

This chapter contains useful legal forms and letters based on your business correspondence with your customers.

CONTRACTS AND AGREEMENTS

Bid for the Purchase of Real Property – Probate

This is a bid for a piece of real estate. It specifies the property to be purchased, the terms, and the method of transfer (warranty deed). Notice that it refers to the seller as the "executor for the estate of the deceased." This is part of this form only because it is very common. You can obviously change it as may become necessary. The bid form also states that you are delivering with it a certified check in the amount of ten percent of the bid amount as a deposit on the property. This is a standard legal form, and should be accepted almost anywhere.

Purchase Agreement

This is a more complex type of sales contract. It specifies far more details about the transfers of goods and money than the previous sales form. In addition, it goes into greater detail on the goods being free of encumbrances and liens, making this a better contract for any property that can be liened. Note the wording of #9: It gives the buyer a time limit for inspection of the goods and filing any damage claim. After that time period, the goods belong to the buyer and may not be returned, regardless of damage. If this paragraph is inappropriate for a certain usage, you can simply remove it and renumber the following paragraph.

Bill of Sale

This is a general bill of sale specifying a one time payment and the unconditional transfer of goods. It is suitable for the transfer of any type of goods.

Bill of Sale – Warranty of Title

This is a bill of sale with warranty of title.

Promissory Note

This is a basic promissory note with interest by which payment is made from one party to another with the second party agreeing to pay the first party back under the terms designated in the note. Notice that the note is payable upon demand after a certain date.

Promissory Note – Installment

This is a more complex promissory note. The terms of this note specify repayment to be made in monthly installments over a predetermined period of time.

Quitclaim Deed

The quitclaim deed is the best method for very quickly transferring real estate from one person to another. Under this agreement, the seller quits all of his or her claims to the property (whether they be income or liability), and transfers them to the buyer. Note carefully that under this agreement it is strictly up to the buyer to ascertain the extent of the assets and liabilities. If the property has liens against it, they are completely transferred to the buyer by the quitclaim deed. Using a quitclaim deed is fast, cheap, and easy. It does, however, subject the buyer to potentially unknown liabilities, and as such, can be financially dangerous.

Release of Deed

This is a standard form for the releasing of a party from deed obligations.

Affidavit

This is a standard affidavit form that can be used for most legal statements. The form leaves an open space for filling in the particulars of the sworn statement. If the statement is long, it may take multiple pages to contain all of the information.

Affidavit of No Lien

This is a sworn and notarized statement of no lien. It is normally used as part of a sale and transfer of real estate or some other type of property. The purpose of this form is to assure the buyer that what they are buying is free and clear of any encumbrances. In order to get a stronger statement of no lien than this, you will normally have to buy title insurance. To this form, you may wish to add a section that reads as follows:

"The seller agrees to hold the buyer free of any possible lien or encumbrance that may have been placed on this property, even if they are unknown to the seller as of this date. If the seller fails in this duty, the buyer will be entitled to whatever compensation and penalties that the law may prescribe, including attorney's fees."

This section guarantees either protection, or whatever types of compensation that the law will allow, should there be any type of lien or encumbrance filed.

Assignment of Deed of Trust

This is a standard assignment of deed of trust form that is used for the sale of real estate. It records the legal title of the property and the deed and transfers the rights that are stated in the agreement.

Assignment of Lease

This is a formal subletting agreement under which one party agrees to assume the lease obligations of another party. Notice that the agreement also requires the consent of the landlord and must be witnessed by third parties.

Assignment of Title

This is a transfer of title from one person, corporation, or partnership to another. Note that this type of assignment of title or right can be used for intellectual property as well as for real property or goods. For example, if you have developed a computer program and wish to assign the rights for it to another party, this form would be better to use than the other types included on this disk.

Assignment of Lien

This is a form for the transfer of lien rights from one person to another in exchange for a cash payment. When using this type of form, make sure that you use care in filling in the many details.

Mechanic's Lien

This is a basic Mechanic's Lien. This document can be used in many places, but you should verify with your lawyer prior to using. Mechanic's Lien laws are not uniform.

Contractor Financial Statement

This is a basic financial statement, but one that is to be signed by the responsible parties. Be careful of these statements as they can be used against you if you report inaccurate numbers.

Certificate of Transaction of Business Under Fictitious Name

This is a fictitious name certification by corporation, partnership, or individual. This is a simple legal notice that is ready to be run in a local newspaper's legal notices section. It is a notice of doing business under a fictitious name for a corporation. Since corporations, partnerships, and individuals must all file fictitious name notices differently, three versions are included here.

Abandonment of Fictitious Business Name

This is a certificate of abandonment of a fictitious name. This is a simple legal notice that is ready to be run in a local newspaper's legal notices section. It simply states that you are abandoning the fictitious name under which you engaged in business. Two versions are included here, one for individuals and one for corporations.

Notice of Right of Rescission

This is notice of rescission, giving a party to a transaction a specified period of time to back out of it. This type of notice is required by the Truth in Lending Act. The seller is required to keep a copy of this notice, so that he or she can prove that the other party actually received the required legal notifications. If the form were not used in the required situations, the seller would be at risk of lawsuits.

LETTERS

Authorization to Negotiate Letter

This is a legal authorization for another person to negotiate on your behalf regarding the named transaction or project. This is a limited type of power of attorney in which the authorized party operates as your agent.

Letter of Credit

This is a cover letter to a purchase order. It reiterates that everything included in the order must be delivered exactly as specified unless the changes are agreed to in writing. To get an extra measure of legal protection, you should include a cover letter like this one with every purchase order that you issue.

Request for Title Letter

This is a request for a preliminary title report on a certain piece of property. In most cases, this type of request comes from a builder/developer who wishes to develop the property. The letter mentions that the appropriate fee is enclosed.

Registration Renewal Letter

Use this as a cover letter for renewal of registration.

Bid for the Purchase of Real Property
(Probate)

To _____ As _____ of the Estate of

_____ Deceased.

The undersigned hereby offers $_____ for the purchase of the property generally known as

_____ and described as:

(description)

on the following terms:

$_____ deposit

(terms for paying balance)

Rents, taxes, insurance expenses of operation, and maintenance to be prorated as of close of escrow;

seller to furnish title policy.

Enclosed is cashier's check #_____ on   (location) Branch of_____(bank) for
$_____ representing 10% of my offer as a deposit.

I understand this bid, if accepted, is subject to confirmation by the Superior Court.

(This offer is contingent on your accepting it by _____ (date) and obtaining the earliest possible date for the court hearing. If you do not accept this offer on or before said date, you are instructed to return the deposit immediately thereafter.) Deposit is to be returned in event of the sale not being confirmed by the court.

Upon confirmation of sale, title shall be conveyed by grant deed (or warranty deed) to _____ (names of parties) (Add description of manner in which title will be held.)

This offer is subject to a real estate commission of ___% to be paid by the estate to (name of broker), _____ (address), License #_____.

Dated: _____   Signature: _____

Address: _____   Phone: _____

ACCEPTANCE

Receipt is acknowledged of $_____ as a deposit. I accept said offer subject to confirmation by the court, and will file a return of sale and immediately notify you of the date of the hearing.

Dated: _____   _____

As executor (administrator)
of Estate of Deceased

As executor (administrator)
of Estate of Deceased

File Type: Word | File Name: 129-Bid-Property-Probate.doc

Purchase Agreement

THIS AGREEMENT, made and entered into this _____ day of _____, 20___, by and between

_____, the Seller, and _____, the Buyer:

1. The seller hereby undertakes to transfer and deliver to the buyer on or before _____, the following described goods:

2. The buyer hereby undertakes to accept the goods and pay for them in accordance with the terms of the contract.

3. It is agreed that identification shall not be deemed to have been made until both the buyer and the seller have agreed that the goods in question are to be appropriated to the performance of the contract with the buyer.

4. The buyer shall make payment for the goods at the time when and at the place where the goods are received by the him/her.

5. Goods shall be deemed received by the buyer when received by him/her at _____ (location).

6. The risk of loss from any casualty to the goods regardless of the cause thereof shall be on the seller until the goods have been accepted by the buyer.

7. The seller warrants that the goods are now free and at the time of delivery shall be free from any security interest or other lien or encumbrance.

8. The seller further warrants that at the time of signing this contract he/she neither knows nor has reason to know of the existence of any outstanding title or claim of title hostile to his/her rights in the goods.

9. The buyer shall have the right to examine the goods on arrival. Within __ business days after such delivery, he/she must give notice to the seller of any claim for damages on account of the condition, quality, or grade of the property. He/She must specify the basis of his/her claim in detail. The failure of the buyer to comply with these rules shall constitute irrevocable acceptance of the goods.

10. Executed in duplicate, one copy of which was delivered to and retained by the buyer, the day and year first above written.

Buyer: _____   Date: _____

Seller: _____   Date: _____

Bill of Sale

I,_____, of (name of firm) , in the County of_____, State of_____, in consideration of _____Dollars, ($____), to be paid by _____, of (name of firm), the receipt of which is hereby acknowledged, do hereby grant, sell, transfer, and deliver unto _____ the following:

_____

_____

(Description)

To have and to hold the same to_____ and his heirs, executors, administrators, successors, and assignees, to their use forever.

And I hereby covenant with the grantee that I am the lawful owner of said goods; that they are free from all encumbrances; that I have good right to sell the same as aforesaid; and that I will warrant and defend the same against the lawful claims and demands of all persons.

In witness, whereof, I_____, hereunto set my hand, this ____ day of_____, 20_____.

_____(Signature)

Seller

Subscribed and sworn to before me this ____ day of _____, 20_____.

NOTARY PUBLIC

_____

My commission expires: _____

Bill of Sale
(With Warranty of Title)

I,_____, of _____ (address) , County of_____, State of _____, in consideration of $_____, paid to me by_____ (seller), the receipt of which is hereby acknowledged, do hereby grant, sell, transfer, and deliver unto _____ (buyer) the following goods and chattels, namely,

(describe items being sold)

To have and to hold the same to _____ and his heirs, executors, administrators, successors, and assignees, to their use forever.

And I hereby covenant with the grantee that I am the lawful owner of said goods; that they are free from all encumbrances; that I have good right to sell the same as aforesaid; and that I will warrant and defend the same against the lawful claims and demands of all persons.

In witness whereof, I _____, hereunto set my hand, this _____ day of _____, 20___.

_____

(Name)

Promissory Note

_____ (city, state, date)

$ (00,000.00)
_____

For value received, the undersigned (jointly and severally, if more than one) promises to pay to _____ the principal sum of _____ dollars, ($00,000.00) with interest from date at the rate of _____ (____) percent per annum on the balance from time to time remaining unpaid. The said principal and interest shall be payable in lawful money of the United States of America at _____ (address) or at such place as may hereafter be designated by written notice from the holder to the maker hereof, on the date and in the manner following: upon demand after _____ (date).

_____
(Name)

_____
(Name)

Promissory Note—Installment

_____ (city, state, date)

For value received, we the undersigned, jointly and severally, promise to pay to the order of _____ (name of lender), _____ (city, state), the sum of _____ ($_____) dollars with interest on any unpaid balance from _____ (date) at the rate of ____ percent per annum, and payable in ____ equal successive monthly installments of _____ dollars in lawful money of the United States of America, commencing on the ____ day of each and every month thereafter until paid, except the final installment which shall be the balance due on this note. If any installment be not paid when due, the undersigned promise to pay collection charges of ____ per dollar of each overdue installment or the actual cost of collection, whichever is greater, and the entire amount owing and unpaid hereunder shall at the election of the holder hereof forthwith become due and payable, and notice of such election is hereby waived. The undersigned promises to pay all reasonable attorney's fees incurred by the holder hereof in enforcing any right or remedy hereunder. All sums remaining unpaid on the agreed or accelerated date of the maturity of the last installment shall thereafter bear interest at the rate of ____ percent per month. The undersigned authorizes the holder to date and complete this note in accordance with the terms of the loan evidenced hereby, to accept additional co-makers, to release co-makers, to change or extend dates of payment, and to grant indulgences all without notice or affecting the obligations of the undersigned, and hereby waives;

    a. Presentment, demand, protest, notice of dishonor, and the notice of nonpayment;

    b. The right, if any, to the benefit or to direct the application of any security hypothecated to the holder until all indebtedness of the maker to the holder, howsoever arising, shall have been paid;

    c. The right to require the holder to proceed against the maker or to pursue any other remedy in the holder's power;

And agrees that the holder may proceed against any of the undersigned, directly and independently of the maker, and that the cessation of the liability of the maker for any reason other than full payment, or any extension, forbearance, change of rate of interest, acceptance, release, substitution of security, or any impairment or suspension of the holder's remedies or rights against the maker, shall not in any way affect the liability of any of the undersigned hereunder. All obligations of the makers, if more than one, shall be joint and several.

_____

(Name)

_____

(Name)

Quitclaim Deed

This quitclaim deed made on _____ (date), between _____(name of transferor), of _____ (address), _____ (city), _____ (county), _____ (state), and _____(name of transferee), of _____ (address), _____ (city), _____ (county), _____ (state).

That for and in the consideration of the sum of _____ (amount) dollars, ($_____), the receipt of which is hereby acknowledged, _____ (name of transferor) does hereby release, remise, and forever quitclaim unto (name of transferee) all of his interest, if any, in that certain real property commonly known as _____ (street address or acreage), located in the City of _____, County of _____, State of _____, described as follows:

(legal description)

Together with all the tenements, hereditament, and appurtenances thereunto belonging and the reversions, remainders, rents, issues, and profits thereof.

To have and to hold, all and singular the premises, with the appurtenances, unto _____ (name of transferee) and his heirs and assigns forever.

In witness whereof, _____ (name of transferor) has hereunto this day and year as set forth above.

_____

(Signature)
(Acknowledgment)

Release of Deed

This release made _____ (date), between _____ , hereinafter referred to as "Releasee", of _____ (address), and _____ , hereinafter referred to as "Releasor", of _____ (address).

Whereas, Releasor desires to settle and has delivered to releasee a deed to the following property:

(legal description)

In and for consideration of this deed, the receipt of which Releasee hereby acknowledges, Releasee hereby releases (reason for giving release).

_____

Date

_____

Signature

File Type: Word | **File Name:** 136-Release-of-Deed.doc

Affidavit

State of _____

County of _____

Before the undersigned, an officer duly commissioned by the laws of _____, on this _____ day of _____, 20___, personally appeared _____ who, having been first duly sworn, deposes and says:

_____

_____

_____

_____

Witness: _____

Sworn and subscribed before me this ___ day of _____ A.D. 20___

_____
Signature

_____
Title

Affidavit of No Lien

STATE OF:_____   COUNTY OF:_____

Before me, a duly commissioned Notary Public within and for the State and County aforesaid, personally appeared_____ _____who, after being duly sworn as required by law, deposes and says:

1. That he is the _____ (title) of _____(firm).

2. That _____ (firm) is the owner of the improved property known and legally described as follows:

    SEE EXHIBIT "A" ATTACHED HERETO

    (or set forth legal description)

3. That _____ is not the subject to any bankruptcy, creditor's reorganization, or insolvency proceeding and none are pending, contemplated, or threatened.

4. That _____ has possession of the property and that there is no other person in possession who has any right in the property.

5. That there are no unrecorded labor, mechanic's, or materialmen's liens against the property and no material has been furnished or labor performed on the property which has not been paid in full.

6. That there are no unrecorded easements or liens of assessments for sanitary sewers, paving, or other public utilities against said property.

7. That there are no claims whatsoever of any kind or description against any fixtures or equipment located on the said premises.

8. That there are no existing contracts for sale, options to purchase, or unrecorded deeds or mortgages existing against said property.

9. That this affidavit is made for the purpose of:

_____

_____

_____
(Signature)
(Name)
(Title)

Subscribed and sworn to before me this ____ day of _____, 20___.

My commission expires_____.

_____
(Notary Public)

Assignment of Deed Of Trust

For value received, the undersigned hereby grants, assigns, and transfers to _____,
_____ all beneficial interest under that certain Deed of Trust
dated _____, executed by _____ Trustor,
to_____ Trustee, and recorded as Instrument No. _____ on
_____ in book _____ page _____, of Official Records in the County
Recorder's office of _____ County, _____ (State), describing land therein as:

(legal description)

Together with the note or notes therein described or referred to the money due and to become due
thereon with interest, and all rights accrued or to accrue under said Deed of Trust.

_____

[Acknowledgment]   _____

_____

Assignment of Lease

_____ and _____
_____, Lessees of that certain lease dated _____, by and
between _____, Lessor and _____, Lessee, and pertaining to that
certain _____ do hereby assign their right, title, and interest
in and to said lease to _____, whose address shall henceforth be _____.
It is agreed and understood that this assignment is contingent upon satisfactory compliance with the
terms and provisions of the lease.

WITNESSES:

_____   _____

_____   _____

ACCEPTANCE

We, _____, hereby accept the above assignment and subject ourselves to all the
promises and covenants therein contained. We fully understand that this assignment is contingent upon
the making of timely payments on the lease and complying with all the terms and provisions of the lease.

WITNESSES:

_____   _____

_____   _____

CONSENT

_____, landlord and lessor under the above lease hereby consents to the
assignment of said lease from _____to _____.

By: _____

Attest: _____

Assignment of Title

This assignment made this _____ day of _____, 20_____, by and between _____
("Assignor"), and _____ ("Assignee"):

Witnesseth that for valuable consideration in hand paid by the Assignee to the Assignor, receipt of which is hereby acknowledged, the Assignor hereby assigns and transfers to the Assignee all of his right, title, and interest in and to all _____ (description) set forth in _____ of that certain _____ Agreement.
Provided, however, no warranties of any kind whatsoever are made incident to this Assignment.

In witness whereof, the Assignor has executed this Assignment on the day and year first above written.

_____ (Signature)

Assignor

_____ (Signature)

Assignee

Subscribed and sworn to before me this ____ day of _____, 20____.

My commission expires_____.

_____

(Notary Public)

Assignment of Lien

This assignment is made on _____(date) by _____ (name of lien holder) _____, of (address) _____, City of _____(city), County of _____ (county), State of _____ (state), herein referred to as "Assignor", to _____ (name of assignee), of _____ (address), City of _____ , County of _____ , State of _____, herein referred to as "Assignee."

In consideration of _____ (amount) dollars, ($_____), receipt of which is hereby acknowledged, assignor does hereby assign to assignee the mechanic's lien on the property of _____(name of property owner), located at _____ (address),_____ (city), _____ (state), which has been duly recorded in the office of (office), in Volume _____ (number), page _____, a copy of which is attached hereto.

Whereas, the intent of this assignment is to transfer to assignee full power to collect that certain sum secured by said lien, assignor does hereby appoint assignee his attorney in fact, with full authority to enforce the lien herein assigned, and to collect and receive the debt secured by said lien, as assignor would do if this assignment were not being made. Any costs incurred by the assignee in enforcing the assigned lien shall be borne by the assignee.

In witness, whereof, assignor has executed this assignment at _____ (place of execution), this ____ (day) of _____ (month) _____, (year)

_____

(Signature)

Mechanic's Lien

Lien Claimant: _____

Property Owner: _____

Property Description: Land and buildings located at _____ (address) in _____ county, of the state of _____, being further described in the _____ County Registry of Deeds in Book _____, Page _____.

AMOUNT DUE:

I, _____, lien claimant, being first duly sworn, claim a lien on the above property for labor performed, said labor furnished with the full knowledge and consent of the owners of said property. The first date of furnishing labor was _____ and the last date was _____.

_____   _____

(Name)   Date
(Title)

STATE OF _____ COUNTY OF _____, _____ (date), 20_____personally appeared the above named _____, and made oath to the truth of the facts contained herein.

Before me, _____

_____

(Notary Public)

Contractor Financial Statement

Date: _____   From: _____

To: _____   Name: _____

_____   Address: _____

_____   _____

Phone: _____   Type of Business: _____

OWNER, PARTNER, OR OFFICER	ADDRESS	TITLE

For the purpose of procuring credit, we furnish the following as a true and correct statement of our financial condition on date named above, and hereby agree to notify you of any materially unfavorable change in our financial condition.

ASSETS		LIABILITIES	
Cash In		Accounts Payable	
Cash on Hand		Notes Payable	
Accounts Receivable		Accruals	
Notes Receivable		Other	
Inventories, Materials, and Supplies			
Other			
		TOTAL CURRENT LIABILITIES	
		MORTGAGES	
TOTAL CURRENT ASSETS		OTHER	
MACHINERY AND EQUIPMENT			
REAL ESTATE		TOTAL LIABILITIES	
PREPAID AND DEFERRED CHARGES		NET WORTH (If Not Incorporated)	
OTHER		CAPITAL STOCK (If Incorporated)	
TOTAL FIXED ASSETS		TOTAL LIABILITIES AND CAPITAL	

REMARKS

_____

Dated at _____ this _____

day of _____ 20_____

_____
(NAME OF CORPORATION, PARTNERSHIP OR INDIVIDUAL)

By_____
(SIGNATURE OF OFFICER, PARTNER, OR INDIVIDUAL)

File Type: Excel | **File Name:** 144-Contractor-Fin-Stmnt.xls

Certificate of Transaction of Business under Fictitious Name

BY CORPORATION:

KNOW ALL MEN BY THESE PRESENTS;

That the undersigned corporation hereby certifies that it is transacting or proposes to transact business in the State of _____, under the fictitious name of _____; and that the principal place of business of said corporation in the State of _____ is located at _____, in the City of _____, County of _____.

_____   _____
(Date)   (Name of Corporation)

[Corporate Seal]   By   _____
                              Corporate Officer

PARTNERSHIP:

IT IS HEREBY CERTIFIED AS FOLLOWS:

1. The undersigned are the partners of a _____ (general or limited) partnership that is transacting or proposes to transact business in the State of _____, under the fictitious name of _____.

2. The principal place of said business is located at _____.

3. The full names and places of residence of the undersigned are: _____, whose place of residence is _____, and _____, whose place of residence is _____.

_____   _____
(Date)   (Name of Partner)

_____   _____
(Date)   (Name of Partner)

INDIVIDUAL:

KNOW ALL MEN BY THESE PRESENTS:

That the undersigned does hereby certify the following:

1. The undersigned is transacting or proposes to transact business in the State of _____ under the fictitious name of _____.

2. The principal place of said business is located at _____.

3. The full name of the undersigned is _____.

4. The place of residence of the undersigned is _____.

_____   _____
(Date)   (Name of Individual)

Abandonment of Fictitious Business Name

PERSONAL:

The undersigned, _____, certifies the following:

1. The undersigned, _____, ceased to use the fictitious name of _____ in transacting business in the State of _____.

2. The full (name or names) and (place or places) of residence of the undersigned _____ (is, are) as follows: _____.

CORPORATION:

The undersigned, _____, certifies the following:

1. The undersigned, _____, ceased to use the fictitious name of _____ in transacting business in the State of _____.

2. The principal place of business of said corporation in the State of _____, is at _____, in the City of _____, County of _____.

3. The above mentioned fictitious name is hereby abandoned.

_____
(Date)

_____
(Signature)

Notice of Right of Rescission

_____

Customer   **Amount**   **Security**

NOTICE TO CUSTOMER REQUIRED BY FEDERAL LAW

Today, _____, you have entered into a transaction which may result in a lien, mortgage, or other security interest on your home. Federal Law provides you with the right to cancel this transaction, if you so desire, without any penalty or obligation at any time within three business days from the above date or the date on which all material disclosures required under the Truth in Lending Act have been given to you. By canceling this transaction, any lien, mortgage, or other security interest on your home resulting from this transaction is automatically void. Any down payment or other consideration you may have tendered on entering this transaction must be refunded to you in the event you cancel. If you desire to cancel this transaction, you may do so by notifying the following party:

_____

_____

_____

by mail or telegram sent by midnight _____, or by any other form of written notice delivered to the above address no later than midnight _____.

Please acknowledge your receipt of this notice by signing the form indicated below.

ACKNOWLEDGMENT OF RECEIPT OF NOTICE

Each of the undersigned hereby acknowledges the receipt of two completed copies of the Notice of Right of Rescission.

_____   Date:_____

_____   Date:_____

Authorization to Negotiate Letter

ABC CONTRACTORS

123 Any Street
Anytown, US 00000
555-555-5555

(Date)

To: _____

This letter will authorize you to negotiate, discuss, and in any other way communicate with _____ in those areas relative to _____.

This letter will further authorize _____ to act in all matters on behalf of
_____

The intent of this authorization is not to be construed to limit, in any way, the power of _____ to act in our behalf, enter into agreements, or contract _____ in both financial areas and sales areas.

Therefore, by the existence of this instrument we hereby authorize _____ to accept or reject agreements, to enter into contracts binding upon _____, and to act in whatever way necessary so as to accomplish that which is being undertaken which may occasion the necessity of this letter.

_____

(Signature)

_____

(Date)

Letter of Credit

ABC CONTRACTORS

123 Any Street
Anytown, US 00000
555-555-5555

(Date)

(Name of Bank)
(Address)
(City, State, Zip)

Dear Sir or Madam:

We hereby agree to accept and pay at maturity any draft or drafts on us, at _____ day's sight, issued by [name of individual] of your city, to the extent of _____ (amount of limit) ($____) dollars, and negotiated through your bank.

Respectfully yours,

_____

(Signature)

Dated: _____

Address: _____

I hereby guarantee the due acceptance of payment of any draft issued in pursuance of the above credit.

_____

Signature of Guarantor

Request for Title Letter

ABC CONTRACTORS

123 Any Street
Anytown, US 00000
555-555-5555

(Date)

(Name)
(Address)
(City, State, Zip)

Dear _____,

Please provide us with a preliminary title report for the following property located in _____ County:

(legal description)

We have enclosed our check made payable to you in the amount of $_____.

Thank you for your prompt attention to this request.

Sincerely,

_____

(Name)
(Title)

Registration Renewal Letter

ABC CONTRACTORS

123 Any Street
Anytown, US 00000
555-555-5555

(Date)

(Name)
(Address)
(City, State, Zip)

Attn: _____

Dear _____,

As per our telephone conversation of the other day, we would like to renew our _____ registration with your office.

Per your request, here are the following items you need:

(detail enclosures here, such as insurance certificates, etc.)

Thank you for your help in this matter. If there is any other problem, please call me at (phone number).

Sincerely,

_____

(Name)
(Title)

CHAPTER SEVEN

Miscellaneous Forms and Letters

This chapter contains miscellaneous forms and letters based on your business correspondence with your customers.

FORMS

Time Management Worksheet

This is a basic time management sheet. The two rules of time management are that you first write down everything important, and second, that you review your records daily. If this form helps you, use it. If not, find something else that will work for you and use that. But, by all means, use something.

Credit Report Authorization

Use this form as authorization for release of credit information. This is a release form that allows the party named to look into your bank or loan account history. This is used as the most common reason for such credit searches, but it can certainly be changed if your circumstances are different.

Managing the Construction Trailer

This is a list of guidelines which should be used for the initial set up and maintenance of construction trailers.

Final Clean Checklist

This is a checklist which is utilized on a site when the job is complete.

Superintendent Attire and Equipment

This is a list for a superintendent outlining the proper attire and equipment needed for a job.

LETTERS

Apology Letter

This is an apology for bad behavior.

Condolence Letter

This is a condolence letter regarding a death. In almost all cases, it is best to briefly express your heartfelt sorrow and sympathy and then to close the letter. This letter also offers your personal help to the bereaved spouse or relative. If you feel deeply about the loss, offer your help and be glad to give it if it is requested.

No Donation Letter

This is a letter for politely declining to donate money to a charity.

Negative Credit Inquiry Letter #1

This is an unfortunate response to a credit referral request.

Negative Credit Inquiry Letter #2

This is another unfortunate response to a credit referral request.

Neutral Credit Inquiry Letter

This letter is based on unknown credit.

Positive Credit Inquiry Letter

Use this for a good credit recommendation.

Information Request Letter

This letter is a denial for information that was requested by a likely hostile party. The letter informs the other party that your attorney has advised you not to release any information unless you are subpoenaed by a court to do so.

Time Management Worksheet

ABC CONTRACTORS

123 Any Street
Anytown, US 00000
555-555-5555

Date: _____   [Mon. Tue. Wed. Thu. Fri. Sat. Sun.]

Time		Appointments	Action Record
6:00 a.m.			
7:00 a.m.			
8:00 a.m.			
9:00 a.m.			
10:00 a.m.			
11:00 a.m.			
Noon			
1:00 p.m.			
2:00 p.m.			
3:00 p.m.			
4:00 p.m.			
5:00 p.m.			
6:00 p.m.			
7:00 p.m.			

Goals for the week:	1. _____
	2. _____
	3. _____
	4. _____

File Type: Excel | **File Name:** 152-Time-Worksheet.xls

Credit Report Authorization

ABC CONTRACTORS

123 Any Street
Anytown, US 00000
555-555-5555

To:_____

RE: Loan #_____ or Savings Account #_____ ,

I hereby authorize release to _____, credit information for my pending credit application on a real estate transaction (or other).

_____
Signature

Loan Opened _____   Monthly Payments _____

High Credit _____   Current Balance _____

Paying Record _____

Savings Account:

Date Opened _____   Present Balance _____

The above is furnished to you in strictest confidence to your request.

Date: _____

By: _____

Managing the Construction Trailer

ABC CONTRACTORS

123 Any Street
Anytown, US 00000
555-555-5555

The Construction Trailer is to be maintained in a clean and well organized manner at all times. This Site Superintendent is responsible for this area.

The following is a list of guidelines to be used in the initial set up of the trailers and will also be used for maintenance.

Exterior
- Sod and landscape must be established and maintained.
- Proper skirting must be installed.
- Steps and landing must be installed and kept clean of debris at all times.
- Signage must be posted including company name and jobsite safety requirements.

Interior
- Carpeted floor must be vacuumed daily. Vinyl floors must be swept daily.
- Filing cabinets must be used properly. (Files are not to be left on desks.)
- Blueprint bins must be properly labeled.
- Job logs must be maintained in the proper desk space area.
- Panel board must always be up-to-date.
- Trash cans must be emptied daily.
- Desk space must be kept clear.

I have read the "Managing the Construction Trailer" requirements and understand the policy.

_____   _____
Signature   Date

File Type: Word | **File Name:** 154-Construction-Trailer.doc

Final Clean Checklist

Project address: _____   Date: _____/_____/_____

City: _____   State: _____   Zip:_____   Inspected by: _____

YES	NO	DESCRIPTION

Entry

1. _____ _____ Address posted.
2. _____ _____ Vinyl and door clean.
3. _____ _____ Doors and door jambs clean.
4. _____ _____ Baseboards clean.
5. _____ _____ Light fixtures clean.
6. _____ _____ Tops of all framework clean.

Living Room

1. _____ _____ Baseboards clean.
2. _____ _____ Room swept/vacuumed.
3. _____ _____ Window stool clean.
4. _____ _____ Windows and window frames clean.
5. _____ _____ Doors clean.
6. _____ _____ Tops of all framework clean.

Kitchen

1. _____ _____ Broiler/oven/range/microwave unpacked and literature placed in cabinet drawer.
2. _____ _____ All protective plastic removed from range and dishwasher.
3. _____ _____ Dishwasher unpacked and literature placed in same drawer with range literature.
4. _____ _____ All cabinets wiped out (inside, doors, trim).
5. _____ _____ Window stools clean.
6. _____ _____ Window and window frames clean.
7. _____ _____ Floors clean.
8. _____ _____ Baseboards clean.
9. _____ _____ Light fixtures clean.

File Type: Word | **File Name:** 155-Final-Clean.doc

Final Clean Checklist (*cont.*)

YES	NO	DESCRIPTION

Master Bedroom

1. _____ _____ Baseboards clean.
2. _____ _____ Room swept/vacuumed.
3. _____ _____ Window stool clean.
4. _____ _____ Windows and window frames clean.
5. _____ _____ Doors and door jambs clean.
6. _____ _____ Tops of all framework clean.
7. _____ _____ Closet floors clean.
8. _____ _____ Closet shelving clean.

Master Bath

1. _____ _____ Window stool clean.
2. _____ _____ Windows and window frames clean.
3. _____ _____ Tub clean.
4. _____ _____ Shower clean.
5. _____ _____ Baseboards clean.
6. _____ _____ Floors clean.
7. _____ _____ Doors and door jambs clean.
8. _____ _____ Tops of all framework clean.
9. _____ _____ Toilet clean.
10. _____ _____ Light fixtures clean.

Hall

1. _____ _____ Baseboards clean.
2. _____ _____ Floor swept/vacuumed.
5. _____ _____ Doors and door jambs clean.
6. _____ _____ Tops of all framework clean.

File Type: Word | **File Name:** 155-Final-Clean.doc

Final Clean Checklist (*cont.*)

YES	NO	DESCRIPTION

Bedroom 2

1. _____ _____ Baseboards clean.
2. _____ _____ Room swept/vacuumed.
3. _____ _____ Window stool clean.
4. _____ _____ Windows and window frames clean.
5. _____ _____ Doors and door jambs clean.
6. _____ _____ Tops of all framework clean.
7. _____ _____ Closet floors clean.
8. _____ _____ Closet shelving clean.

Bedroom 3

1. _____ _____ Baseboards clean.
2. _____ _____ Room swept/vacuumed.
3. _____ _____ Window stool clean.
4. _____ _____ Windows and window frames clean.
5. _____ _____ Doors and door jambs clean.
6. _____ _____ Tops of all framework clean.
7. _____ _____ Closet floors clean.
8. _____ _____ Closet shelving clean.

Bedroom 4

1. _____ _____ Baseboards clean.
2. _____ _____ Room swept/vacuumed.
3. _____ _____ Window stool clean.
4. _____ _____ Windows and window frames clean.
5. _____ _____ Doors and door jambs clean.
6. _____ _____ Tops of all framework clean.
7. _____ _____ Closet floors clean.
8. _____ _____ Closet shelving clean.

Final Clean Checklist (*cont.*)

YES	NO	DESCRIPTION

Additional Room: _____

1. _____	_____	Baseboards clean.
2. _____	_____	Room swept/vacuumed.
3. _____	_____	Window stool clean.
4. _____	_____	Windows and window frames clean.
5. _____	_____	Doors and door jambs clean.
6. _____	_____	Tops of all framework clean.
7. _____	_____	Closet floors clean.
8. _____	_____	Closet shelving clean.

Additional Room: _____

1. _____	_____	Baseboards clean.
2. _____	_____	Room swept/vacuumed.
3. _____	_____	Window stool clean.
4. _____	_____	Windows and window frames clean.
5. _____	_____	Doors and door jambs clean.
6. _____	_____	Tops of all framework clean.
7. _____	_____	Closet floors clean.
8. _____	_____	Closet shelving clean.

Superintendent Attire and Equipment

As a Superintendent, your attire/appearance is a reflection of professionalism as an individual and as a member of this company. Acceptable dress includes a white shirt, conservative jeans (cords, khakis, etc.), and hard-hat. Not only will this type of dress display the type of professional image we see, but it will also help to distinguish Superintendents from trades.

Certain items should be kept with the Superintendent. These include a measuring tape, a hard hat, a notebook, pen/pencil, and cellular phone. Construction can be more efficiently managed if these items are kept on-hand at all times.

Some items of equipment will prove to be necessary to the Superintendent at various times:
- Six foot level – for forms, frame, cornice, and trim punch-outs
- Framing square – for forms, frame, and first trim punch-outs
- Wax Marker – for writing address on inside of front window
- Spray paint – for frame punch-outs
- String line – to check for bows during frame and drywall punch-outs
- Colored adhesive dots – for paint punch-outs
- 100 foot tape – for forms and frame punch-outs

I have read the Attire and Equipment requirements and understand the policy.

_____   _____
Signature   Date

Apology Letter

ABC CONTRACTORS

123 Any Street
Anytown, US 00000
555-555-5555

(Date)

(Name)
(Address)
(City, State, Zip)

Dear _____,

For several days I've done a lot of thinking about my behavior at _____ (time and place). I am writing to offer an apology to you. Please believe that I deeply regret my actions and sincerely ask for your forgiveness. There is no excuse for what I did, but I assure you such behavior will never be repeated.

Respectfully,

_____

(Name)
(Title)

Condolence Letter

ABC CONTRACTORS

123 Any Street
Anytown, US 00000
555-555-5555

(Date)

(Name)
(Address)
(City, State, Zip)

Dear _____,

I was just informed by our office manager of _____'s death. I really don't know how to adequately express my sincere sympathy at this most difficult time, but please accept my deepest condolences.

I had worked very closely with _____ in the office (or other details). Therefore, I feel that I have lost a very good friend as well as a coworker.

Please let me know if I can be of any help to you.

Sincerely,

_____

(Name)
(Title)

No Donation Letter

ABC CONTRACTORS

123 Any Street
Anytown, US 00000
555-555-5555

(Date)

(Name)
(Address)
(City, State, Zip)

Re: _____ (project name or designation)

Dear _____,

Thank you so much for your recent letter requesting a donation to _____. However, it is impossible to support all the worthwhile causes in each area that our company has offices (or other reason). Therefore, our company makes an annual contribution to _____.

Because of the above policy, we are unable to make a contribution to _____. It is a fine project, and we wish you all the best in your endeavors.

Sincerely,

_____

(Name)
(Title)

Negative Credit Inquiry Letter #1

ABC CONTRACTORS

123 Any Street
Anytown, US 00000
555-555-5555

(Date)

(Name)
(Address)
(City, State, Zip)

Dear _____,

Regarding your inquiry as to the credit-worthiness of _____, we regret that we cannot recommend them for credit. Despite the fact that we extended liberal credit privileges to this firm, they failed to pay in a consistent and timely manner. As a result, business between our two companies is now done on a cash basis only.

Sincerely,

_____

(Name)
(Title)

Negative Credit Inquiry Letter #2

ABC CONTRACTORS

123 Any Street
Anytown, US 00000
555-555-5555

(Date)

(Name)
(Address)
(City, State, Zip)

Dear _____,

After careful consideration of your inquiry, we regret that we cannot vouch for the reliability of _____ as a credit risk. Our business dealings with this firm have been unsatisfactory, to say the least. Repeatedly, they have failed to honor their commitment to make payments on time and some of those made were only a partial payment of the amount due. Currently they owe a balance of $_____ for purchases made over _____ months/weeks.

I trust our frankness will be kept in strictest confidence and that we have been of some help to you.

Sincerely,

_____

(Name)
(Title)

Neutral Credit Inquiry Letter

ABC CONTRACTORS

123 Any Street
Anytown, US 00000
555-555-5555

(Date)

(Name)
(Address)
(City, State, Zip)

Dear _____,

In response to your inquiry regarding _____ (company name), we feel we cannot be of much help as our dealings with them have been limited. Generally speaking, our impression of this firm is positive, however, we cannot offer any assurances as to their credit-worthiness.

I am sorry we cannot be more helpful.

Sincerely,

_____

(Name)
(Title)

Positive Credit Inquiry Letter

ABC CONTRACTORS

123 Any Street
Anytown, US 00000
555-555-5555

(Date)

(Name)
(Address)
(City, State, Zip)

Dear _____,

Regarding your inquiry about _____ (subject/name), we are happy to recommend (him/her) to you as an outstanding customer of ours for the past _____ (length of time). Throughout the time we have been associated with _____, he has handled his accounts in an exemplary manner, making every payment exactly as agreed. In addition to being an excellent customer himself, he has recommended others who are equally dependable. It has been our pleasure to do business with _____ (him/her), and we recommend _____(him/her) to you with absolutely no reservations.

Sincerely,

_____

(Name)
(Title)

Information Request Letter

ABC CONTRACTORS

123 Any Street
Anytown, US 00000
555-555-5555

(Date)

(Name)
(Address)
(City, State, Zip)

Dear _____,

Upon the advice of counsel, I must inform you that we are unable to provide you with the information which you requested in your letter of _____.

We will, of course, be happy to provide said information upon service of a subpoena.

We are sorry that we cannot be of further assistance to you presently, and hope that you understand our position in this matter.

Sincerely,

_____

(Name)
(Title)

CHAPTER EIGHT
Glossary

Abstract of title: A written summary of all transactions that could affect the ownership of a piece of real estate including deeds, leases, liens, and wills

Adaptive reuse: Adapting an old or historical building for a new purpose

Addendum (plural: addenda): Written information adding to, clarifying, or modifying a bid. An addendum is generally issued by an owner to a contractor during the bidding process and is intended to become part of the contract

Additive alternate: An alternate bid that, if accepted, adds to the contract sum

Air rights: The right to use the space above a piece of real estate

Air space: A cavity or space in walls, windows, or other enclosed parts of a building between various structural members

Allowance: In contract documents, an amount noted by an architect to be included in the contract sum for a specific item

Alteration: Partial construction work performed within an existing structure; remodeling without a building addition

Alternate bid: The amount to be added to or deducted from a base bid amount if alternate materials and/or methods of construction are required

Anchorage: A secure point of attachment for lifelines, lanyards, or deceleration devices

Anchored bridging: Steel joist bridging connected to a bridging terminus point

Aquifer: An underground formation of sands, gravel, or fractured or porous rock that is saturated with water and that supplies water for wells and springs

Assignment: Transferring the rights and duties under a contract from one party to another

Axial: In a direction parallel to the long axis of a structural member

Backcharge: Billings for work performed or costs incurred by one party that, under the contract, should have been performed or incurred by the party to whom billed

Base bid: An agreed construction sum based on the contract documents

Batter board: Temporary framework used to assist in locating corners when laying a foundation; also used to maintain proper elevations of structures, excavations, and trenches

Bid: A formal offer by a contractor, in accordance with the specifications for a project, to do all or a phase of the work at a certain price in accordance with the terms and conditions stated in the offer

Bid bond: A bond issued on behalf of a contractor that provides assurance to the recipient of the bid that, if the bid is accepted, the contractor will sign the contract and provide a performance bond. The bonding company is obliged to pay the recipient of the bid the difference between the contract's bid and the bid of the next lowest responsible bidder if the bid is accepted and the contractor fails

Bid opening: The actual process of opening and tabulating bids submitted at a prescribed bid date/time and conforming with the bid procedures

Bid security: Funds or a bid bond submitted with a bid as a guarantee to the recipient of the bid that the contractor, if awarded the contract, will accept it

Body belt: A strap with means both for securing it about the waist and for attaching it to a lanyard, lifeline, or deceleration device

Body harness: Straps that may be secured about the person in a manner that distributes the fall-arrest forces over at least the thighs, pelvis, waist, chest, and shoulders with a means for attaching the harness to other components of a personal fall arrest system

Boilerplate: Standardized or formulaic language in a contract

Bolted diagonal bridging: Diagonal bridging that is bolted to a steel joist or joists

Bonding company: A licensed firm willing to execute a surety bond, payable to the owner, securing a contractor's performance on a contract either in whole or in part, or securing payment for labor and materials. Also known as a surety

Breach of contract: A material failure to perform an act required by a contract

Bridging clip: A device that is attached to the steel joist to allow the bolting of the bridging to the steel joist

Bridging terminus: A wall, a beam, tandem joists (with all bridging installed and a horizontal truss in the plane of the top chord), or other element at an end or intermediate point(s) of a line of bridging that provides an anchor point for the steel joist bridging

Builder's risk insurance: Insurance coverage on a construction project during construction

Building code: The legal minimum requirements established or adopted by a government agency for the design and construction of buildings

Building envelope: The outer structure of a building

Bull float: A tool used to spread out and smooth concrete

Bullnose: Any material with a rounded edge, such as a concrete block, ceramic tile, brick, window sill, etc.

Buttress: A projecting structure of masonry or wood to support or give stability to a wall or building against horizontal outward forces

Chase: A groove made in a wall or through a floor to accommodate pipes or ducts

Choker: A wire rope or synthetic fiber rigging assembly that is used to attach a load to a hoisting device

Cleat: A ladder crosspiece of rectangular cross section placed on edge upon which a person may step while ascending or descending a ladder

Cold forming: The process of using press brakes, rolls, or other methods to shape steel into desired cross sections at room temperature

Column: A load-carrying vertical member that is part of the primary skeletal framing system. Columns do not include posts

Competent person: One who is capable of identifying existing and predictable hazards in the surroundings or working conditions which are unsanitary, hazardous, or dangerous to employees and who has authorization to take prompt corrective measures to eliminate them

Connector: A device that is used to couple (connect) parts of a personal fall arrest system or positioning device system together

Connector: An employee who, working with hoisting equipment, is placing and connecting structural members and/or components

Constructibility: The ability to erect structural steel members without having to alter the over-all structural design

Construction load (for joist erection): Any load other than the weight of the employee(s), the joists, and the bridging bundle

Construction manager: An entity that provides construction management services, either as an advisor or as a contractor

Controlled access zone: A work area designated and clearly marked in which certain types of work (such as overhand bricklaying) may take place without the use of conventional fall protection systems guardrail, personal arrest, or safety net to protect the employees working in the zone

Controlled decking zone: (CDZ) An area in which certain work (for example, initial installation and placement of metal decking) may take place without the use of guardrail systems, personal fall arrest systems, fall restraint systems, or safety net systems and where access to the zone is controlled

Controlled load lowering: Lowering a load by means of a mechanical hoist drum device that allows a hoisted load to be lowered with maximum control using the gear train or hydraulic components of the hoist mechanism. Controlled load lowering requires the use of the hoist drive motor, rather than the load hoist brake, to lower the load

Controlling contractor: A prime contractor, general contractor, construction manager, or any other legal entity that has the overall responsibility for the construction of the project—its planning, quality, and completion

Cost breakdown: A detailed summary of all the anticipated costs on a construction project

Critical lift: A lift that (1) exceeds 75 percent of the rated capacity of the crane or derrick, or (2) requires the use of more than one crane or derrick

Deceleration device: Any mechanism such as rope, grab, ripstitch lanyard, specially-woven lanyard, tearing or deforming lanyards, automatic self-retracting lifelines/lanyards

Deceleration distance: The additional vertical distance a falling person travels, excluding lifeline elongation and free fall distance, before stopping, from the point at which a deceleration device begins to operate

Decking hole: A gap or void more than 2 inches (5.1 cm) in its least dimension and less than 12 inches (30.5 cm) in its greatest dimension in a floor, roof, or other walking/working surface. Pre-engineered holes in cellular decking (for wires, cables, etc.) are not included in this definition

Derrick floor: An elevated floor of a building or structure that has been designated to receive hoisted pieces of steel prior to final placement

Double-cleat ladder: A ladder similar in construction to a single-cleat ladder but with a center rail to allow simultaneous two-way traffic for employees ascending or descending

Double connection: An attachment method where the connection point is intended for two pieces of steel which share common bolts on either side of a central piece

Double connection seat: A structural attachment that, during the installation of a double connection, supports the first member while the second member is connected

Effluent: Treated sewage from a septic tank or sewage treatment plant

Equivalent: Alternative designs, materials, or methods that the employer can demonstrate will provide an equal or greater degree of safety for employees than the method or item specified in the standard

Erection bridging: The bolted diagonal bridging that is required to be installed prior to releasing the hoisting cables from the steel joists

Extension trestle ladder: A self-supporting portable ladder, adjustable in length, consisting of a trestle ladder base and a vertically adjustable extension section with a suitable means for locking the ladders together

Failure: Load refusal, breakage, or separation of component parts. Load refusal is the point where the structural members lose their ability to carry the loads

Fall restraint system: A fall protection system that prevents the user from falling any distance. The system is comprised of either a body belt or body harness along with an anchorage, connectors, and other necessary equipment. The other components typically include a lanyard and may also include a lifeline and other devices

Final interior perimeter: The perimeter of a large permanent open space within a building such as an atrium or courtyard. This does not include openings for stairways, elevator shafts, etc.

Fixed ladder: A ladder that cannot be readily moved or carried because it is an integral part of a building or structure. A side-step fixed ladder is a fixed ladder that requires a person getting off at the top to step to the side of the ladder side rails to reach the landing. A through fixed ladder is a fixed ladder that requires a person getting off at the top to step between the side rails of the ladder to reach the landing

Footprint: The outline of a building on the ground, used in site planning

Formwork: The total system of support for freshly placed or partially cured concrete, including the mold or sheeting (form) that is in contact with the concrete as well as all supporting members including shores, reshores, hardware, braces, and related hardware

Girt (in systems-engineered metal buildings): A "Z" or "C" shaped member formed from sheet steel, spanning between primary framing and supporting wall material

Guardrail system: A barrier erected to prevent employees from falling to lower levels

Handrail: A rail used to provide employees with a handhold for support

Headache ball: A weighted hook that is used to attach loads to the hoist load line of the crane

Hoisting equipment: Commercially manufactured lifting equipment designed to lift and position a load of known weight to a location at some known elevation and horizontal distance from the equipment's center of rotation. Hoisting equipment includes but is not limited to cranes, derricks, tower cranes, barge-mounted derricks or cranes, gin poles, and gantry hoist systems. A come-a-long (a mechanical device, usually consisting of a chain or cable attached at each end that is used to facilitate movement of materials through leverage) is not considered hoisting equipment

Hole: A void or gap 2 inches (5.1 centimeters) or more in the least dimension in a floor, roof, or other walking/working surface

Individual-rung/step ladders: Ladders without a side rail or center rail support. Such ladders are made by mounting individual steps or rungs directly to the side or wall of the structure

Jacking operation: Lifting vertically a slab (or group of slabs) from one location to another for example, from the casting location to a temporary (parked) location, or from a temporary location to another temporary location, or to the final location in the structure during a lift-slab construction operation

Job-made ladder: A ladder that is fabricated by employees, typically at the construction site, and is not commercially manufactured. This definition does not apply to any individual-rung/step ladders

Ladder stand: A mobile, fixed-size, self-supporting ladder consisting of a wide, flat-tread ladder in the form of stairs. The assembly may include handrails

Lanyard: A flexible line of rope, wire rope, or strap that generally has a connector at each end for connecting the body belt or body harness to a deceleration device, lifeline, or anchorage

Leading edge: The edge of a floor, roof, or formwork for a floor or other walking/working surface (such as the deck) which changes location as additional floor, roof, decking, or formwork sections are placed, formed, or constructed

Lifeline: A component consisting of a flexible line for connection to an anchorage at one end to hang vertically (vertical lifeline), or for connection to anchorages at both ends to stretch horizontally (horizontal lifeline), and that serves as a means for connecting other components of a personal fall arrest system to the anchorage

Lift slab: A method of concrete construction in which floor and roof slabs are cast on or at ground level and, using jacks, are lifted into position

Limited access zone: An area alongside a masonry wall that is under construction and that is clearly demarcated to limit access by employees

Lower levels: Those areas to which an employee can fall from a stairway or ladder. Such areas include ground levels, floors, roofs, ramps, runways, excavations, pits, tanks, material, water, equipment, and similar surfaces. It does not include the surface from which the employee falls

Low-slope roof: A roof having a slope less than or equal to 4 in 12 (vertical to horizontal)

Maximum intended load: The total load of all employees, equipment, tools, materials, transmitted loads, and other loads anticipated to be applied to a ladder component at any one time

Metal decking: A commercially manufactured, structural grade, cold rolled metal panel formed into a series of parallel ribs; for this subpart, this includes metal floor and roof decks, standing seam metal roofs, other metal roof systems, and other products such as bar gratings, checker plate, expanded metal panels, and similar products. After installation and proper fastening, these decking materials serve a combination of functions including, but not limited to: a structural element designed in combination with the structure to resist, distribute, and transfer loads, stiffen the structure, and provide a diaphragm action; a walking/working surface; a form for concrete slabs; a support for roofing systems; and a finished floor or roof

Multiple lift rigging: A rigging assembly manufactured by wire rope rigging suppliers that facilitates the attachment of up to five independent loads to the hoist rigging of a crane

Nosing: That portion of a tread projecting beyond the face of the riser immediately below

Opening: A gap or void 12 inches (30.5 cm) or more in its least dimension in a floor, roof, or other walking/working surface. For the purposes of this subpart, skylights and smoke domes that do not meet the strength requirements of §1926.754(e)(3) shall be regarded as openings

Opening: A gap or void 30 inches (76 centimeters) or more high and 18 inches (46 centimeters) or more wide in a wall or partition through which employees can fall to a lower level

Permanent floor: A structurally completed floor at any level or elevation (including slab on grade)

Personal fall arrest system: A system used to arrest an employee in a fall from a working level. A personal fall arrest system consists of an anchorage, connectors, and a body harness and may include a lanyard, deceleration device, lifeline, or suitable combination of these. The use of a body belt for fall arrest is prohibited as of January 1, 1998.

Point of access: All areas used by employees for work-related passage from one area or level to another. Such open areas include doorways, passageways, stairway openings, studded walls, and various other permanent or temporary openings used for such travel

Portable ladder: A ladder that can be readily moved or carried

Positioning device system: A body belt or body harness system rigged to allow an employee to be supported on an elevated vertical surface such as a wall, and work with both hands free while leaning backwards

Post: A structural member with a longitudinal axis that is essentially vertical, that: (1) weighs 300 pounds or less and is axially loaded (a load presses down on the top end), or (2) is not axially loaded but is laterally restrained by the above member. Posts typically support stair landings, wall framing, mezzanines, and other substructures

Precast concrete: Concrete members (such as walls, panels, slabs, columns, and beams) that have been formed, cast, and cured prior to final placement in a structure

Project structural engineer of record: The registered, licensed professional responsible for the design of structural steel framing and whose seal appears on the structural contract documents

 File Type: PDF | **File Name:** CF&L Glossary.pdf

Punch list: A list of work that requires correction or completion

Purlin (in systems-engineered metal buildings): A "Z" or "C" shaped member formed from sheet steel spanning between primary framing and supporting roof material

Qualified person: One who, by possession of a recognized degree, certificate, or professional standing, or who by extensive knowledge, training, and experience, has successfully demonstrated the ability to solve or resolve problems relating to the subject matter, the work, or the project

Reshoring: The construction operation in which shoring equipment (also called reshores or reshoring equipment) is placed as the original forms and shores are removed in order to support partially cured concrete and construction loads

Riser height: The vertical distance from the top of a tread to the top of the next higher tread or platform/ landing or the distance from the top of a platform/ landing to the top of the next higher tread or platform/landing

Rope grab: A deceleration device that travels on a lifeline and automatically, by friction, engages the lifeline and locks to arrest a fall

Safety deck attachment: An initial attachment that is used to secure an initially placed sheet of decking to keep proper alignment and bearing with structural support members

Safety-monitoring system: A safety system in which a competent person is responsible for recognizing and warning employees of fall hazards

Self-retracting lifeline/lanyard: A deceleration device containing a drum-wound line which can be slowly extracted from or retracted onto the drum under minimal tension during normal employee movement and which, after onset of a fall, automatically locks the drum and arrests the fall

Shear connector: Headed steel studs, steel bars, steel lugs, and similar devices that are attached to a structural member for the purpose of achieving composite action with concrete

Shore: A supporting member that resists a compressive force imposed by a load

Side-step fixed ladder: See Fixed ladder

Single-cleat ladder: A ladder consisting of a pair of side rails, connected together by cleats, rungs, or steps

Single-rail ladder: A portable ladder with rungs, cleats, or steps mounted on a single rail instead of the normal two rails used on most other ladders

Snaphook: A connector consisting of a hook-shaped member with a normally closed keeper or similar arrangement which may be opened to permit the hook to receive an object and, when released, automatically closes to retain the object

Specifications or specs: A narrative list of materials, methods, model numbers, colors, allowances, and other details that supplements information in blueprints and other working drawings

Spiral stairway: A series of steps attached to a vertical pole and progressing upward in a winding fashion within a cylindrical space

Stair rail system: A vertical barrier erected along the unprotected sides and edges of a stairway to prevent employees from falling to lower levels. The top surface of a stair rail system may also be a handrail

Steel erection: The construction, alteration, or repair of steel buildings, bridges, and other structures, including the installation of metal decking and all planking used during the process of erection

Steel joist: An open web, secondary load-carrying member of 144 feet (43.9 m) or less, designed by the manufacturer, used for the support of floors and roofs. This does not include structural steel trusses or cold-formed joists

Steel joist girder: An open web, primary load-carrying member, designed by the manufacturer, used for the support of floors and roofs. This does not include structural steel trusses

Steel truss: An open web member designed of structural steel components by the project structural engineer of record. For the purposes of this subpart, a steel truss is considered equivalent to a solid web structural member

Steep roof: A roof having a slope greater than 4 in 12 (vertical to horizontal)

Step stool (ladder type): A self-supporting, foldable, portable ladder, nonadjustable in length, 32 inches or less in overall size with flat steps and without a pail shelf, designed to be climbed on the ladder top cap as well as all steps. The side rails may continue above the top cap

Structural steel: A steel member, or a member made of a substitute material (such as, but not limited to, fiberglass, aluminum, or composite members). These members include, but are not limited to, steel joists, joist girders, purlins, columns, beams, trusses, splices, seats, metal decking, girts, and all bridging, and cold-formed metal framing which is integrated with the structural steel framing of a building

Systems-engineered metal building: A metal, field-assembled building system consisting of framing, roof, and wall coverings. Typically, many of these components are cold-formed shapes. These individual parts are fabricated in one or more manufacturing facilities and shipped to the job site for assembly into the final structure. The engineering design of the system is normally the responsibility of the systems-engineered metal building manufacturer

Tank: A container for holding gases, liquids, or solids

Temporary service stairway: A stairway where permanent treads and/or landings are to be filled in at a later date

Through fixed ladder: A fixed ladder that requires a person getting off at the top to step between the side rails of the ladder to reach the landing

Toeboard: A low protective barrier that prevents material and equipment from falling to lower levels and which protects personnel from falling

Tread depth: The horizontal distance from front to back of a tread (excluding nosing, if any)

Tremie: A pipe through which concrete may be deposited under water

Unit price: A predetermined price for a measurement of quantity of work to be performed under a contract. The designated unit price would include all labor, materials, equipment, or services associated with item

Unprotected sides and edges: Any side or edge (except at entrances to points of access) of a stairway where there is no stair rail system or wall 36 inches (.9 m) or more in height, and any side or edge (except at entrances to points of access) of a stairway landing or ladder platform where there is no wall or guardrail system 39 inches (1 m) or more in height

Vertical slip forms: Forms that are jacked vertically during the placement of concrete

Walking/working surface: Any surface, whether horizontal or vertical, on which an employee walks or works, including but not limited to floors, roofs, ramps, bridges, runways, formwork, and concrete reinforcing steel. Does not include ladders, vehicles, or trailers on which employees must be located to perform their work duties

Warning line system: A barrier erected on a roof to warn employees that they are approaching an unprotected roof side or edge and which designates an area in which roofing work may take place without the use of guardrail, body belt, or safety net systems to protect employees in the area

Zoning: Governmental regulations on the use of privately owned land

Titles Available From DeWALT:

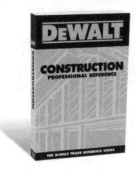

Trade Reference

Blueprint Reading
Construction
Construction Estimating
Construction Safety/OSHA
Datacom
Electric Motor
Electrical Estimating
Electrical Professional Reference
HVAC Estimating

HVAC/R—Master Edition
Lighting & Maintenance
Plumbing
Plumbing Estimating
Residential Remodeling & Repair
Security, Sound & Video
Spanish/English Construction
  Dictionary—Illustrated Edition
Wiring Diagrams

Exam and Certification

Building Contractor's Licensing
Electrical Licensing
HVAC Technician Certification
Plumbing Licensing

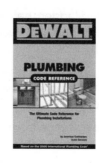

Code Reference

Building
Electrical
HVAC/R
Plumbing

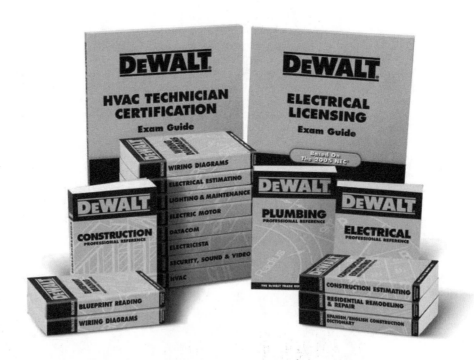

www.DeWALT.com/guides